IMMUNE RESPONSE
AT THE CELLULAR LEVEL

METHODS IN MOLECULAR BIOLOGY

Edited by

ALLEN I. LASKIN
ESSO Research and Engineering
Company
Linden, New Jersey

JEROLD A. LAST
Harvard University
Cambridge, Massachusetts

VOLUME 1: Protein Biosynthesis in Bacterial Systems,
edited by Jerold A. Last and Allen I. Laskin

VOLUME 2: Protein Biosynthesis in Nonbacterial Systems,
edited by Jerold A. Last and Allen I. Laskin

VOLUME 3: Methods in Cyclic Nucleotide Research,
edited by Mark Chasin

VOLUME 4: Nucleic Acid Biosynthesis,
edited by Allen I. Laskin and Jerold A. Last

VOLUME 5: Subcellular Particles, Structures, and Organelles,
edited by Allen I. Laskin and Jerold A. Last
(in preparation)

VOLUME 6: Immune Response at the Cellular Level,
edited by Theodore P. Zacharia

ADDITIONAL VOLUMES IN PREPARATION

IMMUNE RESPONSE
AT THE CELLULAR LEVEL

Edited by THEODORE P. ZACHARIA
National Academy of Sciences
Washington, D.C.

MARCEL DEKKER, INC. New York 1973

MARCEL DEKKER, INC.
95 Madison Avenue, New York, New York 10016

LIBRARY OF CONGRESS CATALOG CARD NUMBER: 73-82621
ISBN: 0-8247-6086-7

PREFACE

In this volume of the series entitled Methods in Molecular
Biology, several expert immunologists were asked to contribute
chapters to the book, each in his own field of specialty. We
all were concerned with the new investigator entering the field
of immunology: the Ph.D., the M.D., the graduate student, the
nurse, the laboratory assistant, the technician, etc. The
authors made every effort to present critically written methods
and sufficient references to give the investigator a comprehen-
sive idea about several areas of immunology: quantitation and
isolation of membrane-associated immunoglobulins and detection
of immunoglobulin secretions by single cells; application of
ferritin-tagged antibodies and immunohistochemical techniques
for isolation of antigens and antibodies; characterization of
antigens of oncogenic viruses; investigation of immunological
tolerance; isolation and purification of complement components;
and isolation of macrophages.

In every chapter, authors present descriptions of methods,
reasons for choosing a specific technique, interpretation of
the results, and references to more detailed works. These references are a valuable nucleus, from which further references
can be obtained.

The investigator is advised to read the companion to this book, also published by Marcel Dekker, Inc., <u>Immunopathology: Methods and Techniques</u>. The two books complement each other and certainly provide the reader with a valuable background in immunology.

Theodore P. Zacharia

LIST OF CONTRIBUTORS

SYDNEY S. BREESE, JR., Plum Island Animal Disease Laboratory, United States Department of Agriculture, Greenport, New York

JACQUES M. CHILLER, Department of Experimental Pathology, Scripps Clinic and Research Foundation, La Jolla, California

B. C. DEL VILLANO, Department of Experimental Pathology, Scripps Clinic and Research Foundation, La Jolla, California

GAIL S. HABICHT, Department of Microbiology, School of Basic Health Sciences, State University of New York at Stony Brook, Stony Brook, New York

KONRAD C. HSU, Department of Microbiology, College of Physicians and Surgeons, New York, New York

S. J. KENNEL, Department of Experimental Pathology, Scripps Clinic and Research Foundation, La Jolla, California

JOHN KLASSEN,* Department of Microbiology, State University of New York at Buffalo, Buffalo, New York

R. A. LERNER, Department of Experimental Pathology, Scripps Clinic and Research Foundation, La Jolla, California

EDWARD J. MOTICKA, Department of Cell Biology, University of Texas, Southwestern Medical School, Dallas, Texas

---

*Present address: Royal Victoria Hospital, Montreal, Quebec, Canada

JOSEPH C. PISANO, Department of Physiology, Tulane University
    School of Medicine, New Orleans, Louisiana

E. SHELTON, National Cancer Institute, National Institutes of
    Health, Bethesda, Maryland

H. S. SHIN, Johns Hopkins School of Medicine, Baltimore, Maryland

R. M. STROUD, Medical College of Alabama, Birmingham, Alabama

WILLIAM O. WEIGLE, Department of Experimental Pathology, Scripps
    Clinic and Research Foundation, La Jolla, California

THEODORE P. ZACHARIA, National Academy of Sciences, Washington,
    D. C.

CONTENTS

PREFACE iii

LIST OF CONTRIBUTORS v

1 APPROACHES TO THE QUANTITATION AND ISOLATION OF
   IMMUNOGLOBULINS ASSOCIATED WITH PLASMA MEMBRANES 1

   S. J. Kennel, B. C. Del Villano, and R. A. Lerner

     I. Introduction 2

    II. Quantitation of Immunoglobulins Associated
         with Plasma Membranes 3

   III. Isolation of Membrane-Associated Immunoglobulin . 13

         Acknowledgments 31

         References 31

2 IMMUNOGLOBULIN SECRETION BY SINGLE CELLS 39

   Edward J. Moticka

     I. Introduction 40

    II. Localized Immune Lysis (Plaque Formation) 41

   III. Isotopic Method (Cluster Technique) 50

    IV. Summary 56

         Acknowledgments 56

         References 56

3   METHODS FOR APPLICATION OF FERRITIN-TAGGED ANTIBODIES . 59

Sydney S. Breese, Jr. and Konrad C. Hsu

    I.   Introduction 60

   II.   Purification and Concentration of Ferritin . . . 61

  III.   Separation of the Globulin Fraction from Serum . 65

   IV.   Conjugation of Ferritin and the Globulin
         Fraction 67

    V.   Evaluation of the Conjugate 70

   VI.   Application of Ferritin Tagging to Tissue
         Cultures 72

  VII.   Experimental Controls in the Application of
         Ferritin-Tagged Antibody 74

 VIII.   Conclusions 75

         References 76

4   IMMUNOHISTOChEMICAL TECHNIQUES FOR THE LOCALIZATION
    OF ANTIGENS AND ANTIBODIES IN CELLS, TISSUES, AND
    ARTIFICIAL SUBSTRATES 79

John Klassen

    I.   Introduction 80

   II.   General Principles 82

  III.   Immunofluorescence Techniques 83

   IV.   Immunoenzyme Technique 95

         References 100

5   CHARACTERIZATION OF ANTIGENS FROM ONCOGENIC VIRUSES . . 103

Theodore P. Zacharia

    I.   Introduction 104

II.   What Influences the Yield of Antigen? 105

III. Source of Antigens 106

IV.   Cells Transformed by RNA Oncogenic Viruses 109

V.   Cells Transformed by DNA Oncogenic Viruses 128

      References 134

6   METHODS FOR THE STUDY OF THE CELLULAR BASIS OF
    IMMUNOLOGICAL TOLERANCE 141

    Gail S. Habicht, Jacques M. Chiller, and
    William O. Weigle

I.   Introduction 142

II.   Methods Applicable to the Study of the
      Cellular Basis of Immunological Tolerance 143

      References 159

7   COMPLEMENT COMPONENTS: ASSAYS, PURIFICATION,
    AND ULTRASTRUCTURE METHODS 161

    R. M. Stroud, H. S. Shin, and E. Shelton

I.   Description of the Complement (C) System 162

II.   Hemolytic Assays 165

III. Purification of Complement Components and
      Other Proteins That Interact with the
      Complement System 177

IV.   Biological Assays 184

V.   Ultrastructure of Clq and Lesions in Membranes
      Produced by the Interaction of Antibody and C:
      Specimen Preparation and Interpretation of
      Results 194

      Acknowledgment 205

      References 205

8   ISOLATION OF MACROPHAGE POPULATIONS 213

Joseph C. Pisano

    I.   Hepatic Macrophages 213

    II. Macrophages of Spleen 215

    III. Pulmonary and Peritoneal Macrophages 217

    IV. Lymphocytes 217

         References 218

AUTHOR INDEX 219

SUBJECT INDEX 231

IMMUNE RESPONSE
AT THE CELLULAR LEVEL

Chapter 1

APPROACHES TO THE QUANTITATION AND ISOLATION
OF IMMUNOGLOBULINS ASSOCIATED WITH PLASMA MEMBRANES*

S. J. Kennel, B. C. Del Villano, and R. A. Lerner

Department of Experimental Pathology
Scripps Clinic and Research Foundation
La Jolla, California

I.   INTRODUCTION 2

II. QUANTITATION OF IMMUNOGLOBULINS ASSOCIATED
     WITH PLASMA MEMBRANES 3

     A. General 3
     B. Selection of the Correct Concentration
         of Antibody 7
     C. Construction of the Standard Curve 8
     D. Measurement of Immunoglobulins
         Associated with Plasma Membranes 11
     E. Interpretation of Results 12

*This is Publication No. 617 from the Department of Experimental
Pathology, Scripps Clinic and Research Foundation, La Jolla,
California. This research was supported in part by grants from
the National Science Foundation, the National Foundation-March
of Dimes, and the Council for Tobacco Research. S. J. Kennel
and B. C. Del Villano were supported by U.S. Public Health Ser-
vice Training Grant 5T1GM683; R. A. Lerner is a recipient of
NIH Career Development Award AI-46372.

III.   ISOLATION OF MEMBRANE-ASSOCIATED IMMUNOGLOBULIN . . . 13

       A. General 13
       B. Lactoperoxidase Iodination 17
       C. Membrane Isolation 23
       D. Dissolution and Separation of Membrane
           Proteins 26

       ACKNOWLEDGMENTS 31

       REFERENCES 31

I.   INTRODUCTION

Gene expression in prokaryotic and eukaryotic cells can be altered by events at cell surfaces. Some examples of this phenomenon are the effects of colicins on bacteria, contact inhibition in mammalian cells, and induction of an immune response. However, the mechanism by which a cell responds to an event at its surface is unknown. In the immune system, the problem can be clearly defined since a single molecular species, the immunogen, induces cellular proliferation in a limited number of reactive cells. Furthermore, recent studies suggest that the union of immunogen with membrane-associated immunoglobulins is the event that initiates cellular proliferation and differentiation [1-3]. In order to understand the immune process and specifically the event(s) at cell surfaces that alter gene expression, it seems logical to study the nature and amount of

membrane-associated immunoglobulins.   In this chapter, two general approaches to the quantitation and isolation of immunoglobulins associated with plasma membranes are detailed.   It should
be noted at the outset that although we are referring to a
specific receptor, i.e., the immunoglobulin receptor, much of
what we have to say is adaptable to the study of other receptors
in diverse biological systems.

II. QUANTITATION OF IMMUNOGLOBULINS ASSOCIATED WITH PLASMA MEMBRANES

A. General

Recently, a modification of the antigen-binding capacity
(ABC) test has been used to quantitate membrane-associated
immunoglobulin and total immunoglobulin in various diploid
human lymphocytes [4-7]. The principle of the ABC test is
quite simple: the reaction between an antibody and a radio-
actively labeled antigen is measured by precipitating the
antigen-antibody complex under conditions where free antigen is
not precipitated.   In the ABC-inhibition technique, unlabeled
antigen is first reacted with a defined amount of antibody, then
labeled antigen is added.   The reaction between the labeled
antigen and remaining unreacted antibodies is measured after
precipitation of antibody-antigen complexes.   By comparison with

the inhibition obtained with known amounts of antigen, the
amount of antigen in a test solution can be determined.

This test may be used to give sensitive and accurate quantitation of any antigen, provided that purified soluble antigen
can be obtained and separated from its corresponding antibody-
antigen complex. For example, we have used this technique to
measure immunoglobulins associated with plasma membranes, as
well as total immunoglobulins, from several cultured human
lymphocytes [5]. In this case, a cell-associated antigen could
be measured because corresponding soluble radioiodinated immunoglobulin fragments were available.

Several points should be made concerning the choice of
reagents. First, the antigen must be pure and soluble--this is
easily obtained with immunoglobulins or immunoglobulin fragments,
but with other antigens this requirement may be a problem. The
antigen is iodinated to achieve about 10,000 cpm per 3.1 ng of
antigen protein. By the chloramine-T method [8], about 500 μg
of antigen protein are labeled with 5 mCi of ^{125}I. After
iodination, the specific activity of the antigen is determined
and a stock solution of the antigen is prepared by dilution of
the labeled antigen to a concentration of 625 ng of protein/ml
in 0.1 M borate buffer, pH 8.3-8.5, containing 2 mg of bovine-
serum albumin per ml (BSA-borate). For immunoglobulin fragments,
the stock solution may be frozen at -20° without loss of antigenic activity. Working antigen solutions are prepared by

dilution of the antigen stock solution to 62.5 ng of protein/ml
in the BSA-borate buffer.

Second, the antibody should be from a hyperimmune serum,
but does not necessarily have to be specific for the antigen,
since the presence of other unrelated antibodies does not interfere with the test. Dilutions of antibody should be made in
10% serum obtained from the same species in which the antibody
was prepared. The diluent used for the determination of immunoglobulins associated with lymphocyte membranes must be chosen
with care to minimize lysis of cells or leakage of internal
immunoglobulins during the test. We have found that Eagle's
minimum essential medium (neutralized with sodium bicarbonate)
is satisfactory. All sera used in this test should have complement removed by heating them to 56° for 30 min.

Finally, two methods may be used for determination of the
amount of antigen-antibody complex formed. (1) A concentration
of ammonium sulfate $[(NH_4)_2SO_4]$ must be found that precipitates
all of the antibody and little of the free antigen. The optimum
concentration of ammonium sulfate depends upon the nature of the
antigen and the species in which the antiserum was prepared.
We have found that the optimum final $(NH_4)_2SO_4$ concentration
for precipitation when rabbit antiserum to human IgG and ^{125}I-
labeled human kappa chain, and $(Fab)_2$ and Fc fragments are studied is 40%, while that for ^{125}I-labeled human λ chain is 44%.
(2) The second method for the quantitation of antigen-antibody

complexes is by precipitation of the complex with a second antibody directed against the antibody present in the complex. This
method, although more cumbersome and expensive, can be used when
the physical properties of the test antigen are such that it cannot be separated from antigen-antibody complexes by precipitation
with $(NH_4)_2SO_4$. Here it is necessary to use an amount of the
second antibody that will completely precipitate the complex.
This method has the disadvantage of requiring larger amounts of
protein or longer incubation time for precipitation than does
the $(NH_4)_2SO_4$ precipitation. This latter technique has been
modified by the addition of ^{131}I-labeled normal IgG and $^{22}Na^+$ to
the reaction mixture [9]. The amount of ^{131}I in the precipitate
gives a measure of the extent of precipitation of the first antibody, while the $^{22}Na^+$ indicates the extent of contamination of
the precipitate with soluble components, i.e., unreacted antigen.
These modifications allow one to recognize and correct "raw"
data for contamination of the precipitate with unreacted
soluble antigen, and for minor technical errors such as incom-
plete removal of supernatants or inadvertent loss of some of the
precipitate. However, a three-channel gamma counter and very
sophisticated computer equipment are required for this modification.

B. Selection of the Correct Concentration of Antibody

Since this assay is based upon a competition between unknown amounts of unlabeled antigen and ^{125}I-labeled antigen for available antibody sites, the correct amount of antibody must be determined. To achieve high sensitivity, we use 6.25 ng of antigen protein for each determination and a dilution of antibody that is capable of binding 50% of the ^{125}I-labeled antigen. Results of a titration of antiserum to human IgG against 6.25 ng of ^{125}I-labeled human kappa chain are shown in Fig. 1. Serial dilutions of antibody are prepared; 100 µl of the ^{125}I-labeled kappa-chain solution (62.5 ng/ml in BSA-borate) is mixed with 100 µl of each dilution of antibody, then incubated at 4° for 1 hr. Then 200 µl of 80% saturated ammonium sulfate is added. Tubes are mixed and incubated at 4° for 30 min, then the precipitate is collected by centrifugation at 1500 x $\underline{g}$ for 30 min at 4°, washed once in 40% $(NH_4)_2SO_4$, and collected by centrifugation as above. For this antiserum, a dilution of 1:750 resulted in the precipitation of 50% of the labeled kappa chain.

C. Construction of the Standard Curve

For estimation of the amount of antigen in a test solution,
it is necessary to determine the ability of the solution to
inhibit the binding of ^{125}I-labeled antigen to a known amount
of antibody. This inhibition is then compared with the inhibition achieved with unlabeled antigen of known concentration.
The detailed procedure for the construction of a standard curve
is as follows:

(i) Inhibitors (i.e., unlabeled antigen of known concentration diluted in BSA-borate): Use 50 µl of inhibitor plus
0.5 ml of diluted antibody. The concentration of inhibitor
depends upon the antigen used, and should cover a wide range.
Tubes are mixed and incubated at 4° for 1 hr.

(ii) Controls

(a) Base-line control (i.e., no inhibitor): Use
50 µl of 10% heated normal serum + 0.5 ml of diluted antiserum
to IgG.

(b) Control for antigenicity of ^{125}I-labeled antigen:
React 100 µl of ^{125}I-labeled antigen with excess antiserum to
IgG (100 µl of a 1:100 dilution). For most preparations more
than 90% of ^{125}I-labeled IgG fragments are precipitated.

(c) Control for nonspecific precipitation of
^{125}I-labeled antigen: React 100 µl of 10% control serum and
100 µl of ^{125}I-labeled antigen.

(iii) From each of the above tubes 100-µl samples are transferred to disposable 12 x 75-mm glass tubes, then 100 µl of ^{125}I-
labeled antigen (6.25 ng/100 µl) is added to each tube. Samples
are mixed and incubated for 1 hr at 4°.

(iv) After the antigen-antibody or control reactions,
ammonium sulfate is added, and after 30 min at 4° the precipitates are collected by centrifugation and washed once with
ammonium sulfate.

(v) The ^{125}I in the precipitate can be determined by any
scintillation counter capable of detecting γ radiation. The
most efficient and cheapest method for detection and/or discrimination between ^{125}I and ^{131}I uses a scintillation counter
containing a NaI crystal (e.g., Packard Instrument Co., LaGrange,
Ill.).

(vi) Calculations

(a) Percent precipitated: The percent of the antigen
precipitated specifically by the antibody is given by

$$\underline{P} = \frac{\underline{A} - \underline{B}(\underline{C} - \underline{A})}{\underline{C}} \text{ X } 100 \tag{1}$$

where $\underline{P}$ = percent precipitated specifically, $\underline{A}$ = the average
cpm precipitated in replicate experimental determinations, $\underline{C}$ =
total cpm added to each tube, and, $\underline{B}$ = the fraction of the total
cpm precipitated in the absence of antibody. It is assumed
that the cpm precipitated nonspecifically is a linear function
of the free cpm in the tube, and, therefore, the factor $\underline{B}(\underline{C} - \underline{A})$

gives the fraction of the unbound antigen that is precipitated
nonspecifically. We have verified the assumption of linearity
of nonspecific precipitation for human kappa chains [9a] but,
even if the assumption is not entirely accurate for other systems, the error in the calculations will be small because the
fraction nonspecifically precipitated is small (usually less
than 10% of the unreacted material). Equation (1) (see above)
can be rearranged, where $\underline{P}$ is the percent precipitated:

$$\underline{P} = \left[\underline{A}\left(\frac{1 + \underline{B}}{\underline{C}}\right) - \underline{B} \right] \text{X } 100 \tag{2}$$

Because this equation has two constants, the percent precipitated is easily calculated on many simple electronic calculators.

(b) Percent inhibition: The percent inhibition is
given by

$$\underline{I}_i = \frac{\left(\underline{P}_o - \underline{P}_i\right)}{\underline{P}_o} \text{X } 100 \tag{3}$$

where $\underline{I}_i$ = percent inhibition for the ith sample, $\underline{P}_o$ = percent
precipitated with no inhibitor and the working antibody dilu-
tion, and $\underline{P}_i$ = the percent precipitated in the ith sample.

We have written a program for the Olivetti Programma 101
to solve these equations (Fig. 2). This program is written for
triplicate determinations for each point, and requires indepen-
dent calculation of $\underline{P}_o$ and $\underline{B}$.

To continue with the kappa chain example, we constructed
a standard curve by plotting the percent inhibition against the

logarithm of the concentration of inhibitor as shown in Fig. 3.
The concentration of antigen in the unknown test solution is
determined by comparison with the standard curve.

D. Measurement of Immunoglobulins Associated with Plasma Membranes

To measure the amount of kappa chain on the surface of
WIL_2 cells, aliquots of washed cells are transferred to test
tubes and sedimented. To each tube, 0.5 ml of antibody is added,
and the cells are mixed and incubated for 1 hr at 4° with
occasional mixing. The cells are then sedimented at 800 x g
for 10 min at 4° and the supernatant is removed. The amount of
unreacted antibody in the supernatant is then determined by
mixing 100 µl of the supernatant with 100 µl (6.25 ng) of the
^{125}I-labeled kappa-chain solution. For 1 x 10^7 WIL_2 cells, the
percent inhibition was 53%, indicating that there is the equivalent of 5.8 ng of kappa-chain protein on the surface of 1 x 10^6
cells. Similarly, the amount of immunoglobulin in the cytoplasm can be measured by mixing 50 µl of appropriate dilutions
of cytoplasm with 0.5 ml of antibody and then continuing as
described above. For the determination of the total immunoglobulin in WIL_2 cells, 5 x 10^7 cells are lysed in
10 mM NaCl, 1.5 mM $MgCl_2$, 10 mM Tris·HCl, pH 7.4 containing
0.5% NP-40. Nuclei are removed by centrifugation at 1000 x g
for 10 min at 4°. Serial threefold dilutions of the supernatant

are made in NaCl-MgCl$_2$-Tris, then 50 µl of each dilution is

used as antigen as described above. The 1:3 dilution showed

56% inhibition, which is equivalent to 190 ng of kappa-chain

protein in the cytoplasm of 10^6 cells.

E. Interpretation of Results

Three principal reservations must be considered in inter-

preting the results obtained by this technique. First, if

inhibition curves are constructed with a single pool of inhibi-

tor immunoglobulin and different hyperimmune anti-IgG prepara-

tions, the inhibition curves may not be identical, but can vary

with the ability of each serum to react with the antigenic

determinants present in the inhibitor immunoglobulin pool.

However, this difference need not lead to erroneous results

for determination of the amount of unknown immunoglobulin,

since reproducible results can be obtained within a system

where a single antiserum and a single inhibitor preparation

are used. Second, comparative data cannot be obtained with

different standard-inhibitor pools. Third, only those antigenic

determinants found on the ^{125}I-labeled antigen can be detected.

This latter point is especially important when myeloma proteins

are used as inhibitor antigens, since each individual myeloma

protein may show restrictions or differences in antigenic

expression. Bearing in mind that only antigenic determinants

that are present in the inhibitor and in the ^{125}I-labeled
antigen, and that are detectable by the antisera, can be measured; the ABC-inhibition technique is sensitive enough to detect
as little as 5 ng of immunoglobulin protein on the surface of
10^6 lymphocytes.

III. ISOLATION OF MEMBRANE-ASSOCIATED IMMUNOGLOBULIN

A. General

The most direct proof of structure is, of course, isolation
and chemical characterization of the molecule. The low concentration of membrane-associated immunoglobulins on lymphocytes
necessitates a sensitive detection method. In addition, the
techniques available for isolation of proteins from membranes
are quite limited and are usually specific for the system in
which they were developed. Thus, while butanol extraction
solubilizes mitochondrial enzymes [10] and chaotropic agents
can break up mitochondrial particles [11] as well as glomerular
basement membrane [11a], neither of these methods is particularly
useful for isolation of membrane-associated immunoglobulins.

Despite these idiosyncrasies, some general methodology can
be applied to the problem. The first consideration is the
choice of cell material. To isolate a homogeneous membrane-

associated immunoglobulin for characterization, one needs a

homogeneous starting material. Tissue culture lines of diploid

human lymphocytes provide a workable system. Our laboratory

has been concerned chiefly with the continuously growing cell

lines 8866 and WIL_2. The molecular biology and growth condi-

tions for these cells have been extensively studied [5, 6, 12,

13, 13a]. Furthermore, clones of these lines can easily be

obtained to satisfy the need for a homogeneous cell population.

The cells are grown in suspension cultures agitated on a gyra-

tory shaker operated at about 100 rpm to keep the cells from

settling out and clumping. In this system, cultures initiated

at 1-2 x 10^5 cells/ml reach a stationary phase of growth at

2-3 x 10^6 cells/ml in about 4 days. Their adaptability to

large culture volumes provides essentially unlimited cell mater-

ial. The reproducible starting material provided by this system

was essential for working out optimum conditions for the proce-

dures outlined below.

The second consideration is that of the concentration of

membrane-associated immunoglobulins. Methods developed in this

laboratory provide an estimate of the concentration of membrane-

associated immunoglobulins on cultured lines (see Section II.D).

The WIL_2 line has about 2 x 10^4 molecules of immunoglobulins/cell

membrane [5]. If one assumes a molecular weight of 160,000,

each cell has about 5 x 10^{-15} g of membrane-associated immuno-

globulin. In more workable numbers this is about 5 ng (see

above) of membrane-associated immunoglobulin protein/10^6 cells.
The total nitrogen content of 10^6 WIL$_2$ cells is about 15 μg.
If we assume that protein is the major N-containing component,
each 10^6 cells has about 100 μg of protein. Thus, by these
estimates the membrane-associated immunoglobulins comprise
about 0.005% of the total cellular protein. The problem is
further complicated since one must distinguish membrane-
associated from intracellular immunoglobulin, which is about
50 times as plentiful in these cells.

In order to detect and quantitate the membrane-associated
immunoglobulin at these low concentrations, a radioactive label
is necessary. The lactoperoxidase system of cell surface iodination supplies this need, as well as a method for discrimination
against proteins not available on the cell surface.

The third consideration is that of membrane isolation.
This procedure is absolutely necessary to show a physical association of the immunoglobulin with membrane fragments. Studies
to date have failed to incorporate this step, and thus cannot
unequivocally discriminate membrane-associated from secreted
immunoglobulin. For example, a molecule in the process of being
secreted at the time of labeling would also be labeled by the
lactoperoxidase procedure. Subsequent procedures would result
in the isolation of secreted, labeled immunoglobulins. The
functional significance of this type of "membrane association"
is questionable. If, however, membranes are isolated after

labeling and shown to be washed free of loosely bound immunoglobulin, the probability of "functional" membrane association
is increased. The membrane isolation procedure used must be
rapid and batchwise, and must give reproducible, high yields
of pure material.

Finally, after membrane isolation, membrane-associated
immunoglobulins must be dissociated and isolated from the membrane. The means of dissociation depends on the nature of the
physical attachment to the membrane. A covalent attachment of
the immunoglobulin to a large membrane structure would present
an enormous problem for chemical isolation, since the nature of
the bond must be determined and selective cleavage methods must
be used. Fortunately, the immunoglobulin-membrane association
is not covalent, but probably hydrophobic in nature. Possible
methods of dissociation include organic solvent extraction,
enzyme treatment, chaotropic ion dissociation, ionic and nonionic detergent treatment, physical disruption, and various
combinations of the above. The method adopted must be strong
enough to dissociate the immunoglobulin, and yet maintain its
antigenic integrity in order to permit isolation.

The usual methods of protein separation, column ion-
exchange chromatography, gel filtration, salt precipitation,
and electrophoresis methods, are not easily adaptable to minute
amounts of protein and often result in large losses of material
during purification. Furthermore, these methods are particularly

unreliable in dealing with membrane proteins. For these reasons, isolation of membrane-associated immunoglobulins by use of immunologic reagents is the most efficient method available.

Two types of isolation can be cited. One type is direct immunologic precipitation. This method is useful for quantitation and characterization as on sodium dodecyl sulfate-polyacrylamide gel electrophoresis [14], but results in immunoglobulin contaminated by the antisera used. A more elegant method of isolation makes use of immunoadsorbent columns. Although technically more difficult, this method can yield a purified product free of any contaminating protein that is suitable for chemical characterization.

Conditions necessary for labeling cells in a particular system may vary. Thus, the procedures by which optimal conditions to maximize yield and purity of the immunoglobulin were derived for the lymphocyte system are presented to aid the reader in adapting this procedure to his specific needs.

B. Lactoperoxidase Iodination

Lactoperoxidase labeling works on the principle that the enzyme is too large to enter the cell. This is true only if the cell membrane is intact and minimal pinocytosis occurs. In addition, the concentration of the reagents involved in the method must be adjusted to give maximum incorporation of I^-

while preserving high-cell viability. The method used for
lymphocytes is a modification of a procedure used for the iodination of erythrocyte membranes [15]. An enzyme concentration
of 0.33 μM was adopted since higher concentrations did not
increase incorporation of label.

A constant carrier iodide concentration of 10 μM KI is
used. This allows the enzyme to function in an efficient substrate range and permits variation of the amount of radioactive
I^- used without affecting the amount of I^- incorporation into
protein. In other words, the proteins are derivitized to the
same extent regardless of the amount of label incorporated.

One critical aspect of the reaction is the amount and time
of H_2O_2 addition. Table 1 shows the results of an experiment
in which H_2O_2 concentration was varied, while ^{125}I incorporation and cell viability were monitored. Maximum incorporation
occurs at a concentration of 44 μM H_2O_2, while cell viability
does not begin to decline until a concentration of 220 μM H_2O_2
is used. In Fig. 4, the kinetics of labeling by this system
are shown. The reaction is quite rapid, so the cells are not
exposed to H_2O_2 and the labeling medium for extended periods
of time. In Fig. 5, the ^{125}I incorporation is shown as a function of the number of additions of H_2O_2. Since incorporation
is not linear with the number of additions, something other
than H_2O_2 must be limiting the reaction.

TABLE 1

Lactoperoxidase[a]--H_2O_2 Concentration

H_2O_2 final conc. (µM)	[125]I incorp. cpm/cell x 10^2	Dead cells %
0	0.2	8
8.8	3.1	7
44	12.7	10
88	10.9	9
220	6.2	17
440	7.8	24

[a]WIL_2 lymphocytes labeled in PBS, 10^{-5} M Kl (10 µC ^{125}I/ml), 3.3 x 10^{-7} M lactoperoxidase, 2 x 10^7 cells/ml (8/16/71).

In the course of these experiments, it was noted that lymphocyte cells did not withstand treatment with phosphate-buffered saline (PBS) (0.15 M NaCl-10 mM PO_4, pH 7.4). Note the relatively poor viabilities in Table 1. For this reason, phosphate-buffered saline was replaced by Earle's balanced salt solution without phenol red. This labeling medium was made 10 µM in KI, adjusted to pH 7.2 with $NaHCO_3$, and used in all subsequent experiments.

The detailed procedure for iodination of the plasma membranes of WIL_2 lymphocytes is as follows.

(i) Stock solutions

(a) Earle's balanced salt solution without phenol red
(Flow Laboratories, Inc., Rockville, Md.) containing 10 μM KI,
adjusted to pH 7.2 with 5.6% NaHCO$_3$ (stored cold after sterile
filtration).

(b) Freshly prepared 0.03% solution of H$_2$O$_2$ in H$_2$O
(8.8 mM). Type 30% (w/w) stock H$_2$O$_2$ should be stored at 4°
shielded from light (H$_2$O$_2$ E$_M$ at 230 nm = 72.4).

(c) Lactoperoxidase, B grade (Calbiochem, Los Angeles,
Calif.) at a concentration of 33 μM (about 6 mg/ml). The stock
solution can be frozen and thawed with only minor loss of activity (E$_{mM}$ at 412 nm = 114).

(d) Suitable dilutions in H$_2$O of Na [125]I (International Chemical & Nuclear Corp., Irvine, Calif.) or Na [131]I
(Cambridge Nuclear Corp., Billerica, Mass.).

(ii) Method: All operations were conducted at 4° except
for the actual labeling, which was done at 25°. Cells were
washed two times in the balanced salt solution containing
10 μM KI and finally adjusted to a density of 2.0 x 10^7 cells/ml.
Lactoperoxidase was added from the stock solution to a final
concentration of 0.33 μM. After addition of an appropriate
amount of radioactive iodide (10-1000 μCi/ml), the reaction
was started by addition of 0.03% H$_2$O$_2$ to a final concentration
of 44 μM. After 1 min, a second addition of H$_2$O$_2$ was made.
One min later, the cells were washed free of enzyme and unreacted iodide. Cells may be washed in any appropriate solution,

but complete growth medium gave better viability. This method resulted in the incorporation of 1-3% of the I⁻ into cellular protein.

Several experiments showed that the iodide was incorporated chiefly into plasma-membrane protein. First, labeled washed cells were swollen in hypotonic solution and disrupted in a Dounce homogenizer. When this suspension was centrifuged at 200,000 x $\underline{g}$ for 1 hr, more than 97% of the acid-precipitable radioactivity was sedimented. This indicates that little if any label is incorporated into intracellular soluble proteins. Second, isolated purified membranes contained more label per µg of N than did the soluble and nuclear fractions. Finally, electron microscopic audioradiographic studies of sections of labeled cells showed grains associated chiefly with the plasma membrane and not with internal membranes or soluble proteins [13a].

TABLE 2

Requirements for Cell Iodination Reaction

Enzyme[a]	H_2O_2 [a]	cpm/10^6 cells	Total incorporation (%)
+	+	665,000	100
+	-	3,400	0.25
-	+	4,300	0.19
-	-	3,800	0.24

[a] +, Presence; -, absence.

This eliminates the possibility that protein was iodinated non-enzymatically after oxidation of I^- to I_2 by H_2O_2. This observation also shows that WIL_2 cells have no complicating endogenous peroxidase activity.

Experiments were done to see if lactoperoxidase was tightly bound to cells and/or pinocytosed. Enzyme was incubated with cells in a complete growth medium either at 37° or at 4° for 15 min. The cells were then washed and subjected to normal labeling conditions without added enzyme. Judging by the amount of enzyme activity found after this procedure, somewhat less than 1% or about 0.5 µg of enzyme/ml remained bound to the cells. Since this occurred at 4° as well as at physiological temperature, nonspecific binding to cells rather than pinocytosis was suspected.

Cells washed only two times before labeling retain a small amount of fetal-calf-serum proteins from the medium bound to their surfaces. These proteins, as well as the added lactoperoxidase, incorporate label during the procedure. Most of these proteins are removed during washings after labeling; however, controls must be included to ensure that this is the case. Specific antisera precipitates of the cell washings after labeling show them to consist chiefly of labeled lactoperoxidase (more than 50% of the label), with some fetal-calf serum (about 20% of label). Two washings remove most of this protein, and the membranes isolated from these cells are relatively free of

contaminating lactoperoxidase and fetal-calf serum (less than
3% of the label).

Cells labeled with 10 µCi of ^{125}I/ml show excellent viability and, in fact, can grow and divide for at least two
generations. At this isotope concentration cells incorporated
about 1 dpm of iodine/50 cells. The specific activity of
labeled membrane proteins was about 10^4 dpm per µg of membrane
nitrogen. This is probably a low estimate since only the proteins on the outside of the membrane are labeled, while the
specific activity is calculated by normalizing total membrane
radioactivity to total membrane protein. Thus, if only 10% of
the proteins in the membrane fraction were actually exposed to
the enzyme during labeling, their specific activity would be
10-fold higher. About 5 x 10^6 atoms of I$^-$ were incorporated per
cell. The amount of incorporated radioactivity varies directly
with the input. At the standard concentration of 10 µCi/ml,
^{125}I is about 5 nM, or about 0.05% of the total unlabeled and
radioactive I$^-$.

C. Membrane Isolation

Several techniques of membrane isolation have been
developed [16]. While these methods are adequate for relatively
small preparations, they are laborious and not easily adapted
to large-scale isolation. The method used in these studies is

that of Brunette and Till [17]. A two-phase aqueous polymer par-

tition [18] is used. Cells are washed twice in 0.16 M NaCl,

then swollen at a density of 5 x 10^7 cells/ml in 1 mM of $ZnCl_2$

for 15-30 min at room temperature. The cells are chilled and

then Dounce-homogenized (B pestle) for 10-70 strokes until

80-90% breakage is attained. All operations are monitored by

phase-contrast microscopy. This is the most irreproducible

step of the procedure and it probably accounts for the varia-

tion in membrane yield and fragment size. The broken cells

are then centrifuged at 1000 x $\underline{g}$ for 15 min and the supernate

(soluble fraction) is removed. The pellet, containing membranes

and nuclei, is suspended in the two-phase system. After cen-

trifugation for 10 min at 15,000 x $\underline{g}$, the two phases are

decanted from the pellet (nuclear fraction) into a clean tube

and mixed. After another centrifugation for 20 min at

15,000 x $\underline{g}$, the membranes are harvested by aspiration from the

two-phase interphase, added to a fresh batch of two-phase

solution, and mixed. After another 10-min centrifugation, the

membranes are aspirated and diluted at least 5-fold in 0.1 M

Tris·HCl pH 8.0 containing 5 mM EDTA (Tris-EDTA). The membranes

are washed in Tris-EDTA and then stored frozen in this solution.

The purified material is a preparation of aggregated plasma

membranes. Membranous material from all isolated fractions--

soluble, nuclear, and membrane--were studied by electron micros-

copy. The purified membrane fraction showed a product similar

to that isolated by other procedures [16].

The cell iodination techniques provide a useful method for monitoring membrane fractionation in this system. Table 3 shows the fractionation of labeled membranes in four such experiments.

TABLE 3

Membrane Distribution in Brunette and Till System [17]

Cell fraction	^{125}I label, %			
	Exp. 1	Exp. 2	Exp. 3	Exp. 4[a]
Membrane	19	23	27	1
Nuclear	20	27	37	54
Soluble	30	19	27	--
Two-phase system	10	15	--	--

[a]Cells labeled in late log phase (9/15/71-10/4/71).

Large fractions of the radioactivity were found in the soluble and nuclear fractions. Electron microscopy showed that plasma membranes were present in these fractions and probably accounted for the presence of ^{125}I.

Table 4 shows the specific activity of the isolated fractions. The membrane fraction is purified considerably by this procedure, while the specific activity of the nuclear fraction is similar to that of whole cells. The soluble fraction, containing the majority of intracellular (nonlabeled) proteins, has a much lower specific activity.

A purified membrane preparation can be made from 10^9 cells in about 3 hr. The method is reproducible and could easily be scaled up to handle 10 times as much material. The washed,

TABLE 4

Nitrogen Distribution in Membrane Preparation

Membrane	μg N/10^6 cells	% N	% ^{125}I	Specific activity cpm/μg N
Whole cells	14.6	100	100	813
Membrane fraction	0.4	2.8	17	5080
Nuclear fraction	4.8	33	39	954
Soluble fraction	6.8	47.5	17	294
Soluble fraction membranes	3.5	24	9	312
Soluble proteins	3.5	24	4	145

labeled membranes provide the starting material for isolation
of a truly membrane-associated immunoglobulin.

D. Dissolution and Separation of Membrane Proteins

In disruption of membrane structure, one must use conditions harsh enough to solubilize membrane proteins and gentle
enough to preserve antigenic properties. To test various
solubilization methods, a suitable concentration of ^{125}I-labeled
membranes was suspended in a solution containing ^{131}I-labeled
immunoglobulin. After the procedure, antisera precipitations
of marker immunoglobulin were conducted to insure that the
method did not destroy antigenic properties of immunoglobulin.

Criteria of solubilization must also be considered.

Membranes must be dissociated such that the pieces do not readily aggregate and precipitate nonspecifically during the precipitin reaction. On the other hand, if membranes are completely dissociated to individual molecular species, no information about proteins surrounding membrane-associated immunoglobulins can be obtained; that is, proteins associated with the immunoglobulin may be of paramount importance in receptor function. If these molecules are dissociated from membrane-associated immunoglobulins by the solubilization procedure, they cannot be isolated by an immunologic method, and information about the environment of membrane-associated immunoglobulins is lost.

Initially, material staying suspended after centrifugation at 10,000 x $\underline{g}$ for 10 min was designated soluble. This procedure selects for structures smaller than about 10^8 daltons. Subsequent precipitin assays showed nonspecific precipitation (trapping) of about 10% of the radioactivity. When criteria for solubility were changed to include only material not sedimenting at 100,000 x $\underline{g}$ for 1 hr (less than about 10^7 daltons), the trapping of nonspecific material dropped to about 1%. The second criterion of solubility thus alleviates the immunological trapping problem, but the yield of solubilized material is somewhat lower.

Most classical dissolution procedures here specified were ineffective in solubilizing the isolated membranes:

(1) solubilization by K-cholate extraction after lipid depletion by acetone extraction [19]; (2) exposure to chaotropes like sodium trichloracetate or potassium iodide [11]; (3) phospholipase digestion; (4) treatment with ionic salts such as Na_2SO_4 or KCl [20]; (5) washing with EDTA; (6) treatment with deoxycholate. Extraction with lithium diiodosalicylate or with the detergent sodium Sarkosyl resulted in good membrane dissolution, but destroyed the antigenic properties of marker immunoglobulin.

The most effective procedure was sonication in the presence of the nonionic detergent NP-40, 0.1 M Tris, and 5 mM EDTA (pH 8.0). Triton X-100 was as effective as NP-40. The material rendered soluble by this treatment is not representative of the distribution of total membrane proteins. The fractions have different protein profiles on sodium dodecyl sulfate-polyacrylamide gel electrophoresis (Fig. 6). Attempts to measure the distribution of membrane-associated immunoglobulin in these two fractions by the antigen-inhibition assay described above were unsuccessful owing to the high detergent concentration. One can detect immunoglobulins in the solubilized fraction, so isolation and characterization can be performed, although exact quantitation is not yet possible.

As mentioned before, two methods of isolation are possible. The first is precipitation by specific antisera. This method permits immunologic characterization of membrane-associated

immunoglobulin, but, owing to contamination with antiserum proteins, does not allow further chemical characterization. The
method involves direct precipitation of the membrane immunoglobulin. If cells are labeled with ^{125}I, subsequent membrane
preparation and solubilization results in a solution containing
labeled membrane immunoglobulin. Twenty μg of ^{131}I-carrier
immunoglobulin of a specific type is added to the preparation.
Addition of the appropriate antiserum should precipitate nearly
100% of the ^{131}I-labeled marker protein, as well as a small
percentage of the ^{125}I-labeled protein. A control for nonspecific trapping [i.e., keyhole limpet hemocyanin (KLH)-anti-
KLH] is done in parallel. Analysis is not complete until the
precipitated proteins are separated on sodium dodecyl sulfate-
acrylamide gel electrophoresis. Several published methods are
adequate for this characterization [21]. The result of one
experiment is shown in Fig. 7. Precipitation with antisera of
different specificities can characterize the membrane-associated
immunoglobulin as a particular type.

Finally, large-scale isolation of membrane immunoglobulin
is possible by antibody immobilized on a solid support. For
this procedure, antiserum is coupled to sepharose 4B by the
cyanogen bromide method [22]. Careful quantitation of column
capacity is necessary. Too low a capacity would soon saturate
with the immunoglobulin, while too high a capacity can cause
complications in the subsequent elution step. An antigen-binding

capacity of about 10 μg/ml of bed volume works well. Care must
be taken to saturate the Sepharose activated with cyanogen bromide with a nonspecific protein before the isolation procedure.

Solubilized ^{125}I-labeled membrane immunoglobulin obtained
as described is diluted in a suitable carrier protein such as
10% fetal-calf serum or 10% normal rabbit serum (provided that
rabbit antibody is used) and run over a column of the antibody-
coupled Sepharose. The yield can easily be monitored by the
use of ^{131}I-marker immunoglobulin. Owing to the solubility
properties of the preparation, NP-40 must be present throughout
this procedure. After absorption is complete, the column is
washed with detergent buffer, then with 1 M NaCl for removal of
nonspecific material. Elution is accomplished with 3 M KSCN
[23]. This reagent breaks antibody-antigen union and allows
the membrane-associated immunoglobulin to pass through the
column. After dialysis against detergent buffer to remove KSCN,
the product must be tested for purity by specific precipitation
and sodium dodecyl sulfate-polyacrylamide gel electrophoresis.
Yields vary from system to system and depend largely on the
quality of the crude solubilized membrane-associated immunoglobulin. Theoretically, this method can provide a purified
immunoglobulin suitable for chemical characterization as to
molecular weight, physical shape and structure, and even
sequence analysis. At present, sequence analysis is not possible owing to the minute quantity of membrane-associated

immunoglobulin available. If we assume a 100% yield from the

starting material, we would need 2×10^{11} cells to produce 1 mg

of immunoglobulin. Thus, actual sequencing of membrane-

associated immunoglobulin awaits more sensitive detection

techniques or a vigorous cell-culture program.

ACKNOWLEDGMENTS

We thank Dr. F. J. Dixon and Miss P. J. McConahey for
their advice and criticism of this work.

REFERENCES

1. S. Sell and P. G. H. Gell, J. Exptl. Med., 122, 423
(1965).
2. C. S. Walters and J. Wigzell, J. Exptl. Med., 132,
1233 (1970).
3. H. Metzger, Ann. Rev. Biochem., 39, 889 (1970).
4. R. Smith, J. Longmire, R. T. Reid, and R. S. Farr,
J. Immunol., 104, 367 (1970).
5. R. A. Lerner, P. J. McConahey, and F. J. Dixon, Science,
173, 60 (1971).
6. R. A. Lerner, P. J. McConahey, I. Jansen, and F. J.
Dixon, J. Exptl. Med., 135, 136 (1972).
7. R. A. Lerner, Contemporary Topics in Immunochemistry
(F. P. Inman, ed.), Plenum Press, New York, 1972, p. 111.
8. P. J. McConahey and F. J. Dixon, Intern. Arch. Allergy,
29, 185 (1966).
9. M. Egan, J. Lautenschleger, J. Coligan, and C. Todd,
Immunochemistry, in press (1972).
9a. B. C. Del Villano, unpublished (1971).
10. R. K. Morton, Methods in Enzymology (S. P. Colowick
and N. O. Kaplan, eds.), Vol. I, Academic Press, New York, 1965, p. 5.
11. Y. Hatefi and W. G. Hanstein, Arch. Biochem. Biophys.,
138, 73 (1970).

11a. H. Marquardt and F. J. Dixon, personal communication.

12. R. A. Lerner and L. D. Hodge, J. Cell. Physiol., 77, (1971) p. 265.

13. R. A. Lerner, W. Meinke, and D. A. Goldstein, Proc. Natl. Acad. Sci., 68, 1212 (1971).

13a. S. J. Kennel and R. A. Lerner, J. Mol. Biol., in press.

14. A. I. Shapiro, F. Vinuela, and J. V. Maizel, Jr., Biochem. Biophys. Res. Commun., 28, 815 (1967).

15. D. R. Phillips and M. Morrison, Biochem. Biophys. Res. Commun., 40, 284 (1970).

16. L. Warren and M. C. Glick, Fundamental Techniques in Virology (K. Habel and N. P. Salzman, eds.), Academic Press, New York, 1969, p. 66.

17. D. M. Brunette and J. F. Till, J. Membrane Biol., 5, 215 (1971).

18. P. Albertson, Adv. Protein Chem., 24, 309 (1970).

19. S. J. Kennel and M. D. Kamen, Biochem. Biophys. Acta, 234, 458 (1971).

20. R. A. Reisfeld, M. A. Pellegrino, and B. D. Kahn, Science, 172, 1134 (1971).

21. J. Weber and M. Osborne, J. Biol. Chem., 244, 4406 (1969).

22. R. Axen, J. Porath, and S. Ernback, Nature, 214, 1302 (1967).

23. W. B. Dandliker, V. A. DeSaussure, and N. Levandoski, Immunochemistry, 5, 357 (1968).

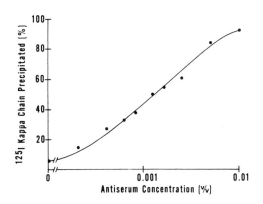

FIG. 1.   Titration of rabbit antiserum to human IgG with
6.25 ng of ^{125}I-labeled human kappa chain.

REG. 1	REG. 2
1 AV	25 D/↓
2 /◇	26 C/−
3 S	27 Bx
4 B/↑	28 D/+
5 S	29 C/÷
6 B ↑	30 E/↑
7 S	31 B/↓
8 C/↑	32 C/−
9 S	33 Bx
10 F ↑	34 B/+
11 S	35 C/÷
12 F/↑	36 C ↕
13 BV	37 C ↓
14 S	38 E/−
15 ↓	39 C÷
16 S	40 E/◇
17 +	41 A ◇
18 S	42 /◇
19 +	43 /◇
20 F/÷	44 CV
21 F−	45
22 A◇	46
23 ↕	47
24 D/↑	48
REG. 1	REG. 2

Constants		Keys to Touch
—		V
P_o		S
B		S
C		S
Background cpm		S
3		S
A_1	triplicate	S
A_2	samples	S
A_3		S

Print out

1) average of A_1, A_2, A_3

2) percent precipitated, P_i

3) percent inhibition, I_i

FIG. 2. Program instructions for Olivetta 101 for the calculation of percent precipitated and percent inhibited in the ABC-inhibition technique.

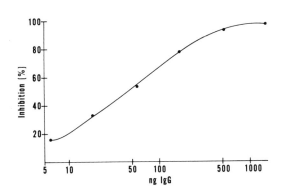

FIG. 3.   Standard curve for the inhibition of precipitation of ^{125}I-labeled kappa chain by human IgG.

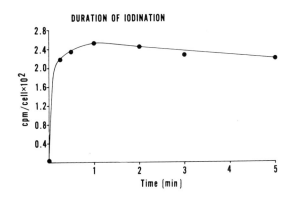

FIG. 4.   Kinetics of incorporation of ^{125}I with lactoperoxidase. WIL2 cells were labeled under the conditions in Table 1, with a final H_2O_2 concentration of 44 µM.

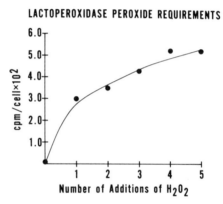

FIG. 5. Lactoperoxidase iodination of WIL_2 cells as in Fig. 4 with H_2O_2 additions every minute.

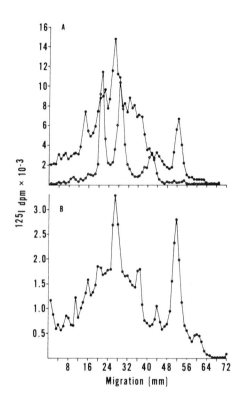

FIG. 6.   Sodium dodecyl sulfate-acrylamide gel electrophoresis.   The ^{125}I-labeled membrane samples were separated in
5% acrylamide gels [21] for 2 hr at 15 mA/gel.   Internal
^{131}I-marker proteins were included in every gel, but shown only
in part A (o——o).   The μ chains, γ chains, and light chains
of human immunoglobulin appear in order of increasing migration.
Part A, total ^{125}I-labeled membrane fraction (●——●); part B,
^{125}I solubilized from total membrane fraction by treatment with
NP40 (●——●).

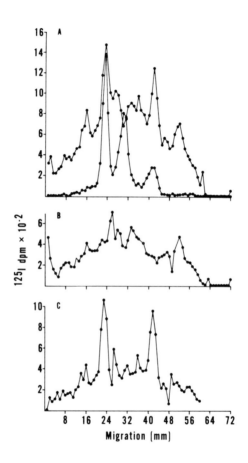

FIG. 7. Sodium dodecyl sulfate-acrylamide gel electrophoresis. Conditions are as in Fig. 6 with the same ^{131}I-marker
proteins (o——o). Part A, solubilized ^{125}I-labeled membrane
proteins precipitated with antiserum to human gammaglobulin
(●——●); part B, ^{125}I-labeled membrane proteins trapped in
an unrelated keyhole limpet hemocyanin (KLH)-anti-KLH precipitate; part C, specific precipitate (A) minus nonspecific precipitate (B).

Chapter 2

IMMUNOGLOBULIN SECRETION BY SINGLE CELLS

Edward J. Moticka

Department of Cell Biology
University of Texas
Southwestern Medical School
Dallas, Texas

I.   INTRODUCTION 40

II. LOCALIZED IMMUNE LYSIS (PLAQUE FORMATION) 41

     A. The Immunodrop Technique 42
     B. Modification to Refine Assay 46
     C. Monolayer Method 49

III. ISOTOPIC METHOD (CLUSTER TECHNIQUE) 50

     A. Materials 52
     B. Procedure 53

IV. SUMMARY 56

     ACKNOWLEDGMENTS 56

     REFERENCES 56

I. INTRODUCTION

Advances in immunobiology during the past decade have been greatly aided by the development of techniques to study the secretion of immunoglobulin molecules by single cells. These methods have enabled immunologists to study more easily the cellular events involved in antibody production. They have also played a major role in providing the answers to several central issues in immunobiology, including the morphology of the immunoglobulin-secreting cell as seen by both light and electron microscopy, the proportion of lymphoid cells within a tissue or organism that can respond to a given antigen, and the specificity of a given immunocyte.

This chapter presents a description of two classes of techniques that have evolved for the study of secretion of immunoglobulins at the single-cell level. The first of these, the technique of localized immune lysis, was described in 1963 and has become a standard instrument in the repertoire of immu- nological methodology. The second procedure, the cluster technique, is a more recent development which appears to have wider applicability because of its increased sensitivity and enhanced scope. While the methods of the first group will detect only cells secreting antibody capable of fixing comple- ment (or made to do so by some subterfuge), the latter tech- niques are theoretically capable of detecting any and all immunoglobulin-secreting cells.

II. LOCALIZED IMMUNE LYSIS (PLAQUE FORMATION)

Nossal [1] originally described a method for detection of the secretion of immunoglobulins by single cells in 1958. His procedure, although elegant, is technically difficult and has not been widely adopted. Five years later, a technique of general applicability was introduced simultaneously by Jerne and Nordin [2] and Ingraham and Bussard [3]. As originally described, the method is able to detect cells secreting antibody to erythrocyte antigens, most usually those on the surface of sheep red blood cells (SRBC). Identification of active cells is based on the ability of the secreted antibody to fix complement and thereby lyse the indicator erythrocytes. Lymphoid cells from an immunized animal are mixed with SRBC in a semi-solid medium. After incubation and the application of complement, clear areas (plaques) appear in an otherwise cloudy layer. The clear areas are caused by lysis of the SRBC, and are visible to the naked eye or with low magnification. Microscopically, a single lymphoid cell can usually be observed in the center of each plaque, and has aptly been termed a "plaque-forming" cell (PFC).

The following two modifications of the original technique have been used in our laboratory; both have similar sensitivities and have been adapted for different purposes. The first modification is well suited for investigations with isotopes, as the preparations are thin and suitable for radioautography.

By the second method results are obtained very rapidly; this
method can be used to screen populations of lymphoid cells for
"antibody producers" while the cells are being prepared for
other manipulations. In this method, plaques appear within 20
min of the start of incubation, yielding a maximum number in
30-45 min.

A. The Immunodrop Technique

This modification was developed by Šterzl and Mandel [4].
Very thin layers of agarose containing suspended lymphoid cells
and SRBC are prepared on microscope slides by dropping the
heated (42°C) mixture from a pipette held at a height of 55 cm.
After fixation and drying, the preparation can be used for
radioautography by any of the conventional techniques.

1. Materials

Materials necessary for this procedure include:

a. 75 x 25 mm microscope slides coated with 1% agarose in
distilled water, dried, and autoclaved

b. 42°C water bath with test-tube rack

c. Ring stand with a pipette holder (or buret clamp)
arranged so that the tip of the pipette is 55 cm above a level
table top.

d. Plastic incubation trays large enough to accommodate microscope slides and constructed with a 0.5-mm space beneath the slide.

e. Hot plate, centrifuge, test tubes, test-tube rack, and bunsen burner

f. 1-ml pipettes with a uniform bore

g. 2-ml and 5-ml volumetric pipettes

h. Glass tissue homogenizers

i. Hemocytometer

j. Sterile gauze

k. Trypan blue; Turck's solution

l. Tissue culture medium, 10-times concentrated--Parker's medium 199 (which is a mixture of just about everything one can think of to feed cells) has been used with success

m. 5% human serum albumin

n. 1% agarose in distilled water

o. SRBC thrice washed in phosphate-buffered saline (pH 7.2)

p. Guinea pig serum (as a source of complement--if chicken lymphoid cells are to be used, chicken serum absorbed three times with SRBC must be used as the complement source since fowl antibody will not fix the C1 component of mammalian complement [5])

2. Procedure

Single-cell suspensions are prepared from the desired

lymphoid organ by gentle manipulation in a tissue homogenizer
in medium supplemented with human serum albumin (HSA). To prepare 100 ml of medium, 10 ml of 10-times concentrated medium
and 10 ml of 5% HSA are added to 80 ml of distilled water. The
pH is adjusted to 7.3-7.4. Cells and medium are kept in an ice
bath throughout the entire procedure until they are plated with
agarose on the slides. Clumps are removed from the cell suspension by allowing it to stand for 5 min in the homogenizer.
The resulting single-cell suspension is poured into test tubes
through sterile gauze, and the cells are washed three times
with excess medium. After the third wash, the cells are diluted
with 1 ml of medium; 0.1 of this is removed, diluted 1:100 in
Turck's solution, and counted in a hemocytometer. With chicken
lymphoid cells, it is better to use the Natt-Herrick stain as a
diluting and counting fluid [6]. To check for cell viability,
0.1 ml of a 0.4% trypan blue solution is added to a tube containing 0.1 ml of the cell suspension in 0.5 ml of medium.
After 5 min (but not more than 15 min), stained (nonviable)
cells are counted in a small drop placed on a slide. From the
results of these two determinations, dilute the remainder of
the cell suspension to provide the proper number of viable cells
for plating. If numerous PFC are expected, it is advantageous
to dilute to a density of between 5×10^5 and 1×10^6 cells/ml.
In situations where few plaques are anticipated, the cells
should be at a density between 5×10^6 and 10^7/ml. It is

generally advisable to run the assay at two or three different
cell dilutions (i.e., 10^5, 10^6, and 10^7 cells per ml).

While the cells are being washed, the agarose-SRBC mixture
can be prepared. Agarose is melted completely, and 4 ml is
pipetted into a previously warmed flask or test tube in a 42°C
water bath with a heated pipette. Then 0.5 ml of the 10-times
concentrated medium and 0.5 ml of 5% HSA are added and mixed
thoroughly. The pH is adjusted, and 1.7 ml of a 4% suspension
of washed SRBC is added. The suspension is divided into 0.4-ml
aliquots into tubes placed in a rack in the water bath. This
protocol is sufficient for 15 or 16 determinations done in
quadruplicate.

With a 1-ml pipette, 0.2 ml of the test cell suspension,
at the proper dilution is removed and mixed with the agarose-SRBC
mixture. The pipette is warmed gently over a flame, and the
contents of the tube are mixed thoroughly and aspirated into
the pipette. The pipette is placed in the buret clamp or
pipette holder and one drop at a time is allowed to fall onto
a coated microscope slide. Three drops can be accommodated per
slide; the contents of a test tube should be sufficient for
four such slides. The drops can be accurately placed on the
slide by testing the apparatus before use and drawing a circle
on the table with a wax pencil where the drop falls.

After the agarose has gelled, the slide is placed, "drop"
side up, into the incubation trays and the trays put in a 37°C

incubator for 2 to 3 hr. Medium may be added to the trays under
the slides to avoid dessication during the incubation period.
At the end of this time, the medium (if any) is discarded, the
slides are inverted, and the trays are flooded with a 10% dilution of the complement source. The slides are then incubated
for an additional 30-45 min. By this time, the plaques are
developed and may be counted at low magnification with dark-
field illumination. Results are usually reported as PFC/10^6
cells or PFC/total cell number of the lymphoid organ tested.

To be prepared for radioautography, these slides are placed
in 3% glutaraldehyde for 30 min, washed for 5 min with running
water, and dried completely at room temperature with an electric
fan. They are then ready for application of the photographic
emulsion.

B. Modifications to Refine Assay

This method, as described, will detect only highly efficient, complement-fixing antibodies, most of which are of the
IgM class. Antibodies of the IgG class can be detected by the
addition of an anti-IgG antiserum in the proper concentration
[7,8]. This antiserum is generally prepared in rabbits and
must be heated at 57°C for 30 min (to eliminate complement
activity), and absorbed three times with SRBC to remove any
naturally occurring anti-SRBC antibodies. To determine the

optimal amount of antiserum to use, a series of control tests
is set up with various dilutions. Most anti-IgG antisera of
moderate to good titer can be used at a dilution of between
1:200 and 1:600. Less dilute preparations may result in the
inhibition of some plaques.

　　To increase the scope of this assay, several antigens can
be conjugated to the SRBC membranes. The resultant "coated"
erythrocytes can then be used to detect antibody-forming cells
of other specificities. Several of the antigens that have been
used and the methods of conjugation are listed in Table 1. One
of the more versatile of the coupling agents is $CrCl_3$. The
technique of coupling SRBC and antigen is quite simple and is
applicable to both proteins and carbohydrates [10]. Equal
volumes of washed, packed SRBC and antigen are mixed in a small
flask. To this mixture is added a similar volume of $CrCl_3$ at
the optimal dilution (usually 0.01 to 1%--this must be empir-
ically determined for each antigen). The reaction is allowed
to proceed at room temperature for 4-5 min with occasional
stirring. It is terminated by addition of an excess of physiological saline; the cells are washed three or four times. If,
at any time during the preparation of the coated erythrocytes,
agglutination is observed, the reaction should be stopped at
once by the addition of excess saline.

TABLE 1

Attachement of Antigens to Erythrocyte Membranes

Antigen	Coupling Agent	Reference
Proteins		
IgG	Rabbit anti-SRBC	13
Serum proteins, myeloma proteins, keyhole limpet hemocyanin, etc.	$CrCl_3$	9
Gammaglobulin, albumin, transferrin, thyroglobulin	Carbodiimide	14
BSA, gammaglobulin	$CrCl_3$	15
Synthetic polymers of amino acids	Tannic acid	16
Carbohydrates		
Somatic antigen of Salmonella enteritidis	NaOH-treated endotoxin	17
Pneumococcal poly-saccharides	$CrCl_3$	10
Lipopolysaccharide	Glutaraldehyde	18
Haptens		
Arsanilic acid	Diazotization	19
Penicillin	Ethylenediaminetetra-acetic acid	20
Various haptens	Chicken anti-SRBC	21

C. Monolayer Method

Cunningham [11] described a method to increase the sensitivity of the plaque technique by dispensing with the gelling
medium (agarose). This was subsequently modified further by
Cunningham and Szenberg [12], and adopted almost in its entirety
for use in our lab.

1. Materials

The materials required to perform this technique are those
for preparation of cell suspensions (i.e., items h-m on the list
under the immunodrop technique, p. 43), SRBC, complement, microscope slides coated with agarose, and double-sided tape (for
example, Scotch brand No. 410). Tape is used in the construction of chambers as follows: a coated microscope slide is
divided into two equal areas by three strips of the double-
sided tape. A second slide is placed over this first one and
pressed into position. If it is desirable to view the PFC under
high magnification, the top slide can be replaced with coverslips. In this case, the slide is divided into three chambers
instead of two.

2. Procedure

Cells washed, counted, and diluted as for the microdrop
technique are mixed with an equal volume of a 4% suspension of

washed SRBC. To this mixture are added complement, to a final concentration of 10%, and any desired developing sera at the appropriate concentration. This mixture is pipetted into the chambers (note the volume, since this will vary) and the chambers are sealed with melted white "Vaseline." After incubation for 30-45 min, the plaques appear and can be counted under low magnification.

The immunodrop and monolayer techniques are comparable in sensitivity, as is indicated in Table 2, which compares the number of PFC detected at various times after SRBC immunization in mouse spleen; the same cell suspension is used in both procedures. The slight increase in sensitivity seen with the monolayer method may result from damage inflicted upon the cells by the short period of exposure to 42°C required for the microdrop procedure. In contrast, both methods showed a large increase in sensitivity when compared with assays performed according to the original technique.

III. ISOTOPIC METHOD (CLUSTER TECHNIQUE)

Detection of immunoglobulin-secreting cells by means of the first two techniques have depended on the antibody activity of the secretion. Recently, Cornille and Rowe [22] have described a method that is capable of detection of immunoglobulin secretion irrespective of its antibody activity. Similar

TABLE 2

Comparison of the Immunodrop and Monolayer Techniques
for Detection of Anti-SRBC Antibody-Secreting Cells

Day after immunization	Direct PFC/10⁶ spleen cells		Indirect PFC/10⁶ spleen cells	
	Monolayer	Immunodrop	Monolayer	Immunodrops
First immunization				
0 (22)[a]	0.6	0.2	(6)[a] 0	0
4 (8)	740	400	(8) ND[b]	ND
7 (5)	190	127	(5) ND	ND
11 (5)	52	55	(5) 407	209
Second immunization				
5 (3)	127	87	(3) 1790	1170
6 (4)	102	146	(4) 1380	1204
11 (2)	15	17	(2) 287	181

[a]All values in a given horizontal row were obtained from the same cell suspension. Numbers in parentheses indicate the number of mice killed on each day.
[b]ND = not determined.

techniques [23,24] had been used only for the detection of
secretors of specific antibody. The following modification of
the Cornille and Rowe procedure has proved quite sensitive and
applicable to investigation of cells secreting both specific
antibody and nonspecific immunoglobulins.

The principle of the cluster technique is similar to that
of the plaque method. Lymphoid cells are incubated in a semi-
solid medium containing labeled antiimmunoglobulin instead of
indicator erythrocytes. After undergoing incubation, washing,
fixing, drying, and exposure to photographic emulsion, the cells
secreting immunoglobulin can be visualized by the clusters of
grains around their periphery.

A. Materials

Materials needed are identical to those listed for the
immunodrop technique, except that ^{125}I-labeled antiimmunoglobulin
is substituted for SRBC. Antiserum to immunoglobulin is usually
produced in rabbits and purified on specific immunoabsorbent
columns according to the method of Rejnek et al. [25]. Labeling
with ^{125}I is accomplished according to the method of McConahey
and Dixon [26]. Optimal labeling conditions are obtained by the
addition of 0.25 ml of chloramine-T (2 mg/ml), 0.5 ml of antiserum (2-3 mg/ml) and 0.25 ml of ^{125}I (carrier-free, 10 mCi/ml).
The reaction is allowed to proceed at room temperature for 10 min

with stirring, and 0.5 mg of Na_2SO_3 (2 mg/ml) is added to the

mixture to stop the reaction. The products are subsequently

passed through a Sephadex G-25 column and dialyzed overnight

against phosphate-buffered saline to remove unbound iodine.

This produces ^{125}I-labeled antiimmunoglobulin possessing a

specific activity of 500 µCi/mg.

B. Procedure

For the usual assay, 0.2 ml of a washed single-cell sus-

pension (1-5 x 10^6 cells/ml) is added to 0.2 ml of agarose (made

in medium as for the immunodrop technique) and 0.2 ml radioac-

tive medium. The radioactive medium is prepared by addition of

0.25 ml of the ^{125}I-labeled antiimmunoglobulin antiserum to

25 ml of medium supplemented with HSA, streptomycin (200 µg/ml),

penicillin (200 U/ml), and sodium azide, at a final concentra-

tion of 2 mg/ml. This mixture, maintained at 42°C in a water

bath, is taken up with a 1-ml pipette and deposited one drop at

a time from a height of 55 cm onto microscope slides coated with

agarose. When the agarose has set, the slides are inverted into

incubation trays which are subsequently flooded with incubation

medium. This medium is prepared by dilution of the excess

labeled medium used in the preparation of the slides 2:1 with

unlabeled medium containing HSA, penicillin, streptomycin, and

sodium azide as above. Thus, the concentration of the

^{125}I-labeled antiimmunoglobulin is identical on the slides and
in the incubation medium. After incubation at 37°C for 2-4 hr,
the slides are washed in phosphate-buffered saline at 4°C for
2 days with frequent changes. After fixation in 3% glutaraldehyde for 30 min, being washed in running water for 5 min and
dried, the slides are ready for any radioautographic assay.

When this method is used, it is always necessary to distinguish between those cells actively secreting immunoglobulins
and those that merely have immunoglobulins on their surface.
This differentiation can be accomplished by running a control
with the same suspension whereby the cells are exposed to the
label under conditions (incubation at 4°C) at which secretion
will not occur. As indicated in Table 3, different lymphoid
organs show different percentages of labeled cells with the two
temperatures. The difference between the two sets of values is
an indication of the minimum number of cells in the suspension
that are actively secreting immunoglobulins. Although the percentage of cells labeled in the spleens of different animals
will vary owing to numerous factors, including strain and immunological condition, the difference between the two techniques
is fairly constant among animals subjected to the same treatment
(Moticka, unpublished observations).

TABLE 3

Percentage of Cells in Various Lymphoid Organs of Mice
That are Labeled with ^{125}I-Labeled Antiimmunoglobulin[a]

Labeling conditions	Spleen	Lymph node	Peripheral blood	Bone marrow	Peyer's patches	Thymus
37°C	46.8	27.3	36.2	8.9	54.7	0.22
4°C	28.6	22.7	15.6	2.7	52.4	0.03
Difference[b]	18.2	4.6	20.6	6.2	2.3	0.19

[a]Data when the reaction is performed at either 4° or 37°C.
[b]The difference in the percentage of cells labeled by the two methods is assumed to represent the minimum percentage of cells that actively secrete immunoglobulin (IgG).

IV. SUMMARY

In preliminary experiments the immunodrop and cluster techniques have been combined so that the secretion of specific
antibody and nonspecific immunoglobulin could be studied on
the same slide. At low concentrations of antiimmunoglobulin
(to overcome the inhibitory effects of high concentrations of
antiimmunoglobulin on formation of plaques), it has been
observed that all PFC are cluster-forming cells, but not all
cluster-forming cells are PFC. This combination of methods is
being used to investigate the underlying cellular correlates
of tolerance, enhancement, and antigenic competition.

ACKNOWLEDGMENTS

Much of the original work on these methods was performed
in the laboratory of Dr. Jaroslav Šterzl during the tenure of
a fellowship from the National Academy of Sciences (USA). I
thank Dr. Šterzl for generously providing research facilities,
Dr. Petr Šima for assistance with the radioautography, and
P. Dana Johanovská for technical advice.

REFERENCES

1. G. J. V. Nossal, Brit. J. Exptl. Pathol., 39, 544-551
(1958).
2. N. K. Jerne and A. A. Nordin, Science, 140, 405 (1963).

3. J. S. Ingraham and A. Bussard, J. Exptl. Med., 119, 667-684 (1964).

4. J. Šterzl and L. Mandel, Folia Microbiol. 9, 173-176 (1964).

5. H. N. Benson, H. P. Brumfield, and B. S. Pomeroy, J. Immunol., 87, 616-622 (1961).

6. M. P. Natt and C. A. Herrick, Poultry Sci., 31, 735-738 (1952).

7. J. Šterzl and I. Říha, Nature, 208, 585-859 (1965).

8. D. W. Dresser and H. H. Wortis, Nature, 208, 859-861 (1965).

9. E. R. Gold and H. H. Fudenberg, J. Immunol., 99, 859-866 (1967).

10. P. J. Baker, P. W. Stashak, and B. Prescott, Appl. Microbiol., 17, 422-426 (1969).

11. A. J. Cunningham, Nature, 207, 1106-1107 (1965).

12. A. J. Cunningham and A. Szenberg, Immunology, 14, 599-600 (1968).

13. M. Segre and D. Segre, J. Immunol., 99, 867-875 (1967).

14. E. S. Golub, R. I. Mishell, W. O. Weigle, and R. W. Dutton, J. Immunol., 100, 133-137 (1968).

15. G. H. Sweet and F. L. Welborn, J. Immunol., 106, 1407-1410 (1971).

16. P. Walsh, P. Maurer, and M. Egan, J. Immunol., 98, 344-350 (1967).

17. M. Landy, R. P. Sanderson, and A. L. Jackson, J. Exptl. Med., 122, 483-504 (1965).

18. M. Eskenazy and B. Petrunov, Immunology, 21, 311-312 (1971).

19. B. Merchant and T. Hraba, Science, 152, 1378-1369 (1966).

20. B. E. Harrell and B. Merchant, Intern. Arch. Allergy, 32, 21-26 (1967).

21. H. Silver, J. F. A. P. Miller, and N. L. Warner, Intern. Arch. Allergy, 40, 540-550 (1971).

22. R. Cornille and D. S. Rowe, Clin. Exptl. Immunol., 8, 981-986 (1971).

23. E. Pick and J. D. Feldman, Science, 156, 964-966 (1967).

24. N. R. Klinman and R. B. Taylor, Clin. Exptl. Immunol., 4, 477-487 (1969).

25. J. Rejnek, R. G. Mage, and R. A. Reisfeld, J. Immunol., 102, 638-646 (1969).

26. P. J. McCohahey and F. Dixon, Intern. Arch. Allergy, 29, 185-189 (1966).

Chapter 3

METHODS FOR APPLICATION OF FERRITIN-TAGGED ANTIBODIES

Sydney S. Breese, Jr.

Plum Island Animal Disease Laboratory
ARS, U.S. Department of Agriculture
Greenport, New York

and

Konrad C. Hsu

Department of Microbiology
College of Physicians and Surgeons
New York, New York

I. INTRODUCTION 60

II. PURIFICATION AND CONCENTRATION OF FERRITIN 61

III. SEPARATION OF THE GLOBULIN FRACTION FROM SERUM . . . 65

IV. CONJUGATION OF FERRITIN AND THE GLOBULIN
FRACTION 67

V. EVALUATION OF THE CONJUGATE 70

VI. APPLICATION OF FERRITIN TAGGING TO TISSUE
CULTURES 72

VII.   EXPERIMENTAL CONTROLS IN THE APPLICATION
       OF FERRITIN-TAGGED ANTIBODY 74

VIII. CONCLUSIONS 75

       REFERENCES 76

I. INTRODUCTION

Immunological reactions may be studied by use of the electron microscope with ferritin-tagged antibodies. The technique, which is somewhat time-consuming owing to the preparation of reagents, provides an easy marker for investigation of the reaction between antigens in tissues or cells and antibodies.

Electron microscopic techniques and details of the preparation of ferritin and the conjugation to antibody will be given in this chapter. A few ways in which ferritin is applied to cellular or tissue culture systems will also be discussed. Two recent articles give some idea of the scope of the immunoferritin technique in various biological systems [1,2].

Methods for the preparation of ferritin and its conjugation with antibody are given in such a way that the technique should be within the capability of any reasonably well-equipped laboratory. The details are important and must be adhered to very carefully. As with all relatively complicated procedures, the new investigator will always be aided by the opportunity to visit a laboratory in which immunoferritin techniques are being

used. A short apprenticeship, in other words, will supplement
any set of directions.

II. PURIFICATION AND CONCENTRATION OF FERRITIN

Ferritin solutions are currently available from many sources of biochemical supplies, and are generally in a relatively
crude (crystallized twice) or in a more purified (crystallized
five or six times) form. All suppliers use cadmium sulfate
precipitation since it provides the form of ferritin that can
be carried through to conjugation. In general, it has been our
experience that all ferritin supplied commercially must be precipitated again and purified immediately before use. The general
directions that follow are for a preparation of horse-spleen
ferritin containing about 1 g in 10-11 ml of solution that has
been crystallized only twice (Pentex, for example).

The chemical solutions required are reagent-grade cadmium
sulfate in distilled water (20% w/v), ammonium sulfate in distilled water (2% w/v, pH 5.85), and ammonium sulfate in distilled
water (saturated, more than 80 g/100 ml).

A phosphate buffer (0.05 M, pH 7.5) is also used; it is
prepared with distilled water and stock solutions of 0.1 M
Na_2HPO_4 (eight parts) and 0.1 M KH_2PO_4 (two parts). This buffer is used frequently and can be made in large quantities.
General laboratory equipment such as a refrigerated centrifuge,

sterilizing (Millipore) filters, dialysis tubing, an untracentri-
fuge (Spinco, Beckman Instruments, Palo Alto, California), a
light microscope, and slides are also required.

The protocol followed is designed to remove the portion of
the ferritin that crystallizes after treatment with cadmium sulfate. The original ferritin solution is diluted to a concentration of 10 mg/ml with 2% ammonium sulfate. When the product from
Pentex (Kankakee, Ill.) is used, the 1 gram per bottle is diluted
in 100 ml of 2% ammonium sulfate. Then 33.3 ml of 20% cadmium
sulfate solution is added, to a final concentration of 5% cadmium
sulfate. The pH of the 2% ammonium sulfate is critical; it must
be 5.85. The mixture is stored overnight at 4-6°C. The bottles
are then centrifuged for 2 hr at 1060 x g in an International
PR-2 refrigerated centrifuge. The supernate, which contains the
nonprecipitable ferritin, is decanted and the resulting sediment
of ferritin crystals is diluted again to 100 ml with 2% ammonium
sulfate. A second centrifugation performed as previously is followed by decantation into clean bottles; the sediment, which contains extraneous protein, is discarded. The ferritin in solution
is recrystallized by the addition of 33.3 ml of 20% cadmium sulfate and storage at 4° overnight. After storage, a small portion
of the crystals are removed with a small pipette and examined
under a light microscope. At this stage, there should be some
crystals that have a yellow-orange color and a characteristic
hexagonal double-tetrahedron structure. There will also be

considerable amorphous residue. Recrystallization may have to be repeated two or three times to result in a field that shows only crystals. The size of the crystals will vary, but a true hexagonal star shape will be unique and characteristic of the correct product. Conversion to apoferritin may occur when the pH of the 2% ammonium sulfate is not correct. The pH must be measured accurately with a meter and adjusted to precisely 5.85, but no higher than 5.88. Some 0.1 M NaOH or 0.1 M HCl is added very slowly to adjust the pH.

The next steps are designed to exchange cadmium ions for ammonium ions. The final crystals from the preceding steps, formed after centrifugation of the cadmium sulfate precipitate and removal of the fluid, are dissolved in 75 ml of 2% ammonium sulfate. Then, an equal volume (75 ml) of saturated ammonium sulfate is added, resulting in a 50%-saturated solution. This solution is allowed to stand for a few hours at 4-6°C, and is centrifuged at 1060 x g for 2 hr. The resulting supernatant fluid is discarded. This second time, the crystals are dissolved in 75 ml of distilled water. Then, 75 ml of saturated ammonium sulfate is added and the mixture is stored at 4°. Centrifugation and another precipitation from distilled water of the crystals should be sufficient to remove all the cadmium ions. It is very important to be sure the ammonium sulfate is saturated at room temperature before it is used. The high amount of salt used (80 g/100 ml or more) is absolutely necessary

because an unsaturated solution will lead to difficulties in
recovery of ferritin in the dialysis steps that follow.

The final precipitate is dissolved in a minimal amount of
distilled water, no more than 3-4 ml per bottle. Careful and
gentle resuspension allows complete recovery, and also a high
concentration of final product. The precipitate is washed gently
out of the bottle and placed in a dialysis tube that will allow
for a 100% increase in volume. The bag is put in a cylinder and
allowed to dialyze against cold (14°-16°C), running tap water
overnight. This step is followed by dialysis against several
thousand ml of 0.05 M phosphate buffer (pH 7.5) at 4°-6°C overnight. A slowly moving magnetic stirrer is desirable to keep
the phosphate buffer moving around the dialysis bag.

The resulting solution is recovered, centrifuged at low
speed to remove any extraneous solids, passed through a Millipore
filter, and stored in a small sterile bottle until needed for
conjugation. At this stage, the ferritin concentration should
be between 30 and 45 mg/ml. Concentration measurements at this
and other stages of the procedures are most easily made with a
hand-held protein refractometer (Hitachi-Perkin Elmer, Norwalk,
Connecticut). When it is necessary to concentrate ferritin, it
is spun in a Spinco 40 or 40.2 rotor at 40,000 rpm for 3 hr at
a temperature of 10°C. The top 3/4 of the colorless solution
is removed from the resulting pellet, and 1-2 ml of 0.05 M
(pH 7.5) phosphate buffer is added. The tubes are then allowed

to stand covered in a refrigerator overnight, by which time most
of the pellet will be resuspended. Any coagulated solids in the
bottom of the tubes may be removed by light centrifugation. The
highly concentrated ferritin is again passed through a Millipore
filter and stored in a sterile bottle or vial. Ferritin concentration may be as high as 65-75 mg/ml. Ferritin and its conjugates may be stored for a long time when sterile techniques are
used during storing and handling.

III. SEPARATION OF THE GLOBULIN FRACTION FROM SERUM

Preparation of ferritin is only half of the procedure
leading to conjugation. For the most specific, and presumably
most reactive, ferritin conjugates the globulin fraction of
serum is used. Conjugation to whole serum is also used in some
cases and has been successful with certain viruses. The separation of globulin is a standard procedure, given here as a con-
venience. Any suitable method may be used, but these directions
are for sodium sulfate precipitation.

The solutions needed include 0.85% (w/v) of sterile sodium
chloride in distilled water, reagent-grade anhydrous sodium sulfate with a solubility at 25°C of not less than 4.8 g in 100 ml
of distilled water, and 0.1 M phosphate-buffered saline (pH 7.2)
made up in sterile distilled water from 85 ml of 0.25 M Na_2HPO_4,
15 ml of 0.25 M KH_2PO_4, and 25 g of NaCl.

Sterile procedures are used throughout these preparations
to prevent contamination of the serum and globulin. Dialysis
tubing is boiled in distilled water, handled with sterile forceps, and stored in a sterile petri dish. Sterile glassware
and pipettes are also used to minimize the accumulation of
extraneous proteins, since the procedure takes several days.

Serum should be sterile when used; it is diluted with an
equal volume of 0.85% NaCl solution. For each 10 ml of diluted
serum, 2.04 g of anhydrous sodium sulfate is added very slowly,
with very gentle stirring to prevent foaming. This solution is
allowed to stand covered at room temperature (25°C) for at least
2 hr, then centrifuged for 30 min at 1060 x g in an International
PR-2 centrifuge at 25°. The supernate is discarded and the
precipitate is dissolved in sterile distilled water to a volume
equal to two-thirds of the volume of the original diluted serum.
A second precipitation is performed by the addition of 1.93 g
of anhydrous sodium sulfate for each 10 ml of distilled water
that was added. This precipitation takes place at room temperature (25°C) for about 2 hr or more. This step is followed by
centrifugation and another precipitation with 1.93 g of sodium
sulfate per 10 ml of distilled water. After centrifugation the
supernate should be colorless and may be discarded. The precipitate is taken up in a minimum volume of sterile distilled water.
With several rinses of the containers, there should not be more
than 1.5-2.0 ml of distilled water for each 10 ml of undiluted

original serum. This procedure should allow transfer to a
dialysis bag without completely dissolving the precipitate.
The small amount of water is necessary to maintain high concentrations for later procedures.

The precipitate is allowed to dialyze against cold running
water (14°-16°C) until it is salt-free. The precipitate first
becomes cloudy until the volume increases and the solution
clears. Be sure to allow for a 100% increase in volume in the
dialysis bag. Reprecipitation occurs when the solution becomes
salt-free, and may occur anywhere from 30 min to 4 hr after the
start of dialysis. This procedure should result in concentrations of globulin of 20-30 mg/ml. Solutions are passed through
a Millipore filter and stored in sterile containers. The sodium
sulfate procedure should work for all sera that have a low lipid
content and contain no more than 8% protein (w/v).

IV. CONJUGATION OF FERRITIN AND THE GLOBULIN FRACTION

Once purified ferritin and globulin have been prepared,
the two may be brought together with a crosslinking agent to
produce ferritin-tagged antibody. The methods are quite simple,
but take about 48 hr; thus, it is often best to prepare them
just before a weekend. Ferritin is first attached to the
crosslinking chemical, metaxylene diisocyanate. The mixture

is then added to the serum globulin and the second step of the process takes place. The final product has combining sites that will react with antigen and produce the desired localization in the sample for electron microscopy.

In addition to the purified ferritin and purified globulin, the following solutions will be needed: (1) sodium borate buffer (0.3 M, pH 9.5) prepared by titration of 0.3 M boric acid with 0.3 M NaOH to pH 9.5 (the use of sodium borate crystals does not result in a stable solution); (2) phosphate buffer (0.05 M; pH 7.5) as was previously described; (3) ammonium carbonate (0.1 M) in distilled water.

Metaxylene diisocyanate (MXDI) is purchased from a biochemical supply house (Polysciences, Inc., Warrington, Pa.) in 1-ml ampules. The clear liquid is transferred to small capillary tubes (0.05-0.1 ml), which are sealed by melting at both ends. These capillaries are stored in a refrigerator until used. Any visible white crystals in the tubes are an indication of oxidation, and they must be discarded. When metaxylene diisocyanate is mixed with ferritin and the serum globulin is added, only glass "fleas" may be used as stirring bars. Teflon stirrers react with the solution and prevent attachment. Glass "fleas" may be made by sealing short lengths of small iron nails in Pyrex tubing.

The reactions take place in solutions adjusted to a final concentration of 0.1 M borate. In the original mixture, for example, we want to have a final concentration of 20-25 mg/ml

of ferritin in 0.1 M borate. Therefore, we might take 2 ml of

ferritin at a concentration of 75 mg/ml and add 2 ml of phos-

phate buffer and 2 ml of 0.3 M borate buffer. This will result

in a total volume of 6 ml containing 0.1 M borate and 25 mg of

ferritin/ml. This mixture is put into a small (25 ml) Erlen-

meyer flask with a glass "flea." Then 0.1 ml of MXDI is added

for each 100 mg of ferritin. In the example cited, that means

0.15 ml because we used a total of 150 mg of ferritin. This

mixture is stirred slowly with the flask in an ice bath for

45 min. The contents are then transferred to a conical centri-

fuge tube and spun at 1000 x g for 30 min. The resulting sedi-

ment is a yellowish sludge. The clear supernate is removed and

saved for the next step. All glassware, the flea, and the cen-

trifuge tube should be washed immediately to remove any MXDI

residue, as it hardens rapidly and is then difficult to remove.

For the next reaction mixture we assume that in the previ-

ous procedure we recovered 5 ml that contained 125 mg of ferri-

tin. We add, for example, 1 ml of globulin (containing 30 mg

of protein) and 0.4 ml of 0.3-M borate, and the resulting solu-

tion contains about 0.1 M borate; the ratio of globulin to fer-

ritin protein is 1:4.

The reaction mixture is placed in a 25-ml Erlenmeyer flask

with a glass "flea" and mixed gently at 4-6°C for 48 hr. The

solution is then dialyzed overnight at 4° against 1 liter or more of 0.1-M ammonium carbonate. This step is followed by

dialysis against 0.05 M phosphate buffer for another 24 hr.

The mixture is removed from the dialysis bag and centrifuged

for 30 min at 1000 x $\underline{g}$. The supernate is then centrifuged for

4 or more hr at 40,000 rpm in a Spinco ultracentrifuge. The

pellets are resuspended as before by being allowed to dissolve

in 0.05 M phosphate buffer (only 2 ml are put in each tube).

The final preparation is then sterilized by passage through a

Millipore filter and stored in a sterile container. The conju-

gate is then ready for use and may be held sterile in a refrig-

erator for a year or more. The total preparation time from the

beginning of this procedure may be almost two weeks.

V. EVALUATION OF THE CONJUGATE

After preparation of the ferritin conjugate, it is usually

assessed for its efficacy. The reaction between an antigen and

a combining site on the conjugate must be completed before the

reaction can be visualized in the electron microscope. Several

methods may be used to determine how the conjugate will react

in an experimental situation. For those laboratories that are

so equipped, immunoelectrophoresis on glass slides may be used

for a rapid answer. The wells may be filled with antigens,

while the trough may have ferritin conjugate and rabbit antiserum

to ferritin, a mixture that will show the reaction to both the
ferritin and the globulin portions of the conjugate. A somewhat simpler method may be the use of agar-gel diffusion in
petri dishes. The basic diffusion medium contains 1% washed
agar, 1.0 M glycine, and 12.5 mM barbital adjusted to pH 7.8-7.9
with 0.1 M HCl. Also 0.5% sodium azide is added, as a preserva-
tive. We use 15 ml of melted agar in an 80-mm plastic petri
dish. The plates are incubated at room temperature in a humidified chamber, and the precipitation reactions are allowed to
develop 2-4 days before being photographed. Very often results
can be seen in 24 hr [3]. By use of the ferritin conjugate in
the center well and a pattern of six wells around the periphery
with two-fold dilutions of antigen, a rough estimate may be made
as to the concentration that may be effective in an actual
experiment. The reverse may also be done, with antigen in the
center well and two-fold dilutions of conjugate on the periphery.
Either procedure will show one or two concentrations where good
precipitation lines have formed midway between the wells. While
the distinctive color of ferritin needs no confirmation of its
presence, rabbit antiserum to ferritin, which is commercially
available, may also be used to form a precipitate in agar-gel
diffusion. If the two-fold dilutions--say 1/2-1/32--are all
reactive, a second gel diffusion with 1/10, 1/20, 1/30, etc.
dilutions would allow a better assessment of the precipitin
endpoint.

VI. APPLICATION OF FERRITIN TAGGING TO TISSUE CULTURES

Examples of how ferritin-tagged antibodies may be used are given throughout the literature of biochemistry and histochemistry. The methods used specifically in the direct and indirect procedures to test for a virus in tissue culture cells [4,5] will be outlined here. The details are not very different from those of when ferritin conjugates are used to study cell surfaces (see Chapter 2) or those of reactions in small blocks of tissue.

As shown in Fig. 1, the direct and indirect tagging methods require the same starting materials. In investigations of animal disease the viral antiserum is usually made in cattle, pigs, or sheep. In the direct test, this serum is then fractionated to remove globulin, which is conjugated to ferritin in accordance with the methods described above. In the indirect method, on the other hand, the globulin fraction in the viral antiserum is used to make antiserum to itself in some appropriate host, such as a rabbit, goat, or guinea pig. The antibody to the globulin of the host species of the antiviral antiserum is then conjugated to ferritin.

In a study of equine arteritis virus in tissue cultures of equine dermis cells [6], infected cells were removed from prescription bottles, washed twice with Sorenson's phosphate buffer (pH 7.2), and pelleted gently in a table-top centrifuge in a conical glass tube. In direct tagging experiments, these cells were gently resuspended in 0.05-0.1 ml of equine antibody to arteritis virus that is conjugated to ferritin. In some cases equine serum was conjugated and used, while in other experiments the globulin fraction was used. The suspended cells were held at room temperature for 20-30 min, centrifuged gently, washed twice with phosphate buffer to remove unreacted ferritin, then fixed in glutaraldehyde. This sequence was followed by osmium fixation and standard embedding in Epon.

In the indirect method, the infected cell pellet was treated with the globulin fraction of the equine antiserum to the virus and allowed to stand at room temperature for 20-30 min. This step was followed by two or three washes with phosphate buffer to remove the free antibody. The cells were gently pelleted again, resuspended in the ferritin conjugate (0.05-0.1 ml), and allowed to stand at room temperature for 20-30 min. After this procedure, the pellets were washed two or three times with phosphate buffer, fixed, and embedded for electron microscopy. The indirect method uses a ferritin conjugate to the antibody to the globulin of the host species of the antiviral antibody. Since this antibody has reacted with virus particles in the cells, there may be more than one antibody-combining site available on the virus-antibody complex to which the ferritin conjugate may attach. For this reason, the indirect method may be more sensitive than the direct method. It also has the advantage of requiring a smaller number of conjugations. We maintain

stocks of rabbit antibovine serum, rabbit antiporcine serum,

and rabbit antiequine serum, all attached to ferritin. When a

diagnostic sample is checked, the antigen and antibody of the

appropriate species are used and the corresponding ferritin is

added.

VII.   EXPERIMENTAL CONTROLS IN THE APPLICATION OF
FERRITIN-TAGGED ANTIBODY

When experiments are done with ferritin-tagged antibodies,

it must be ascertained that nonspecific reactions do not inter-

fere with interpretations of the results. The primary aim in

the use of ferritin methods is to identify the morphology or

the site of the antigen. Therefore, to be effective, antibody

must be made that is as monovalent as possible. The antigen

used to produce the antibody to which the ferritin will be con-

jugated must be as pure as possible. When tissue cultures or

cell suspensions are used, the antibody must be free of nonspe-

cific reacting material. In some cases it is necessary to

absorb the antiserum with other antigens before conjugation.

Some samples should be treated with unconjugated antibody;

others should receive ferritin conjugate alone, and non-infected

cells must be included in the controls.

Examination in an electron microscope must include all

areas where antigen and antibody may combine. The final deci-
sion as to whether a particular ferritin conjugate has reacted
positively or negatively is made after a comparison of the sites
in all the samples. Some samples are expected to be positive,
while others are expected to be negative. Variations are used
not only to determine the efficiency of the antigen-antibody
reaction, but also to serve as checks on the removal of unreacted reagents by washing, on the quality of the cell and antigen
preservation, and on the embedding and sectioning.

VIII. CONCLUSIONS

The use of ferritin-tagged antibody is only one method
that may be used in conjunction with electron microscopy to
study immunological reactions. There are other methods, such
as labeling with peroxidase, the use of hybrid antibodies that
have specificities to two different antigens (one of which is
electron visible), and more complex techniques (such as autoradiography). The use of ferritin tagging is a very positive
method that immediately shows in an electron microscope the
location of the reaction. It is the easiest technique for the
new investigator to master, and it offers many applications to
immunological problems.

REFERENCES

1. C. Howe, C. Morgan, and K. C. Hsu, in Progress in Medical Virology, Vol. II, Karger, Basel, New York, 1969.

2. G. A. Andres, K. C. Hsu, and B. C. Segal, Handbook of Experimental Immunology, Vol. 2 Cellular Immunology (D. H. Wier, ed.), 2nd ed., Blackwell Scientific Publications, Oxford, England. Chapter 33 (1973).

3. K. M. Cowan, J. Immunol., 97, 647 (1966).

4. S. S. Breese, Jr., J. Gen. Virol., 4, 343 (1969).

5. S. S. Breese, Jr., J. Gen. Virol., 8, 153 (1970).

6. S. S. Breese, Jr., and W. H. McCollum, Arch. ges. Virusforsch., 35, 290 (1971).

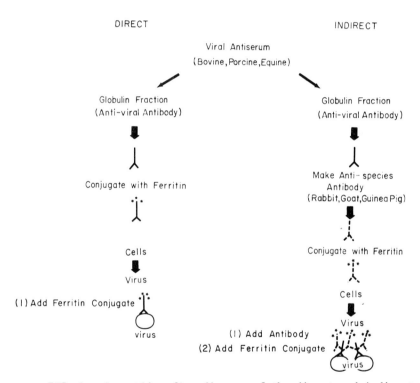

FIG. 1.   An outline flow diagram of the direct and indirect methods of ferritin tagging, as applied to a virus-cell system.

Chapter 4

IMMUNOHISTOCHEMICAL TECHNIQUES FOR THE LOCALIZATION OF ANTIGENS
AND ANTIBODIES IN CELLS, TISSUES, AND ARTIFICIAL SUBSTRATES

John Klassen*

Department of Microbiology
State University of New York
Buffalo, New York

I.   INTRODUCTION 80

II. GENERAL PRINCIPLES 82

     A. Antibody 82
     B. Antigen 82

III. IMMUNOFLUORESCENCE TECHNIQUES 83

     A. Optics 83
     B. Labeling with Fluorochromes 85
     C. Processing Specimens 88
     D. Staining Procedures 89
     E. Tests for Specificity 91
     F. Photomicrography 94
     G. Applications of the Immunofluorescence Technique . 94

IV. IMMUNOENZYME TECHNIQUE 95

     REFERENCES 100

*Present address: Royal Victoria Hospital, Montreal, Quebec,
Canada.

I. INTRODUCTION

Numerous histochemical methods have been devised for the
identification of structures in cells and in tissues by light
microscopy [1]. Several dyes have been used to stain tissue
components with different colors. Advantage has also been taken
of the property of some substances to fluoresce on exposure to
light of certain wavelengths. For example, changes in ultraviolet fluorescence of proteins are a very sensitive index of
conformational changes in cells [2]. In these studies, as well
as in other areas of histochemistry, increasing use is being
made of sensitive photomultipliers and recording devices to
quantitate the staining [3]. Identification of chromosomes
recently has been made relatively easy by the discovery that
quinacrine hydrochloride binds to the DNA of chromosomes and
also fluoresces under ultraviolet light [4]. Fluorescein diacetate, which easily penetrates mammalian cell membranes, is
rapidly hydrolyzed in the cells. As long as the cell is viable
free fluorescein stays in the cytoplasm, but it rapidly leaks
out of dead cells. This property is the basis of a lymphocyto-
toxicity assay in which the number of living cells can be easily
counted because they are all fluorescent [5]. Cellular enzymes
have been localized in tissues by production of variously colored
reaction products with appropriate substrates [1,6]. Since some
of the reaction products are electron dense, enzymes have been

localized also by electron microscopy [6,7]. Enzymes, as well

as ferritin, which is also electron dense, have been used by

several investigators as trace substances in the body; for

example, they have been used to study the permeability of the

glomerular basement membrane with the aid of the electron micro-

scope [8,9]. Autoradiography with radioactive elements such as

^{125}I and ^3H has also been used in light- and electron-microscopy

studies to localize substances in tissues, as well as to demon-

strate the migration of lymphocytes in the body [10].

The exquisite specificity of antibody for its corresponding

antigen forms the basis of the immunohistochemical techniques.

Since an appropriately labeled antibody will react only with

its antigen, very accurate localization can be achieved. Fluo-

rescent compounds such as fluorescein isothiocyanate and tetra-

methyl rhodamine isothiocyanate [10-12], enzymes such as horse-

radish peroxidase, phosphatases, and cytochrome C [6], and

electron-dense substances such as ferritin, uranium, and viruses

[14,15] have been used to label antibodies and antigens. Since

enzymes also have great specificity for their substrates, an

analogous system has been worked out to reveal the localization

of various substrates with enzymes labeled with fluorochromes

or with ferritin [16]. In this chapter, some of the more gen-

eral principles and some significant recent advances in the

immunofluorescence and the immunoenzyme techniques will be

discussed.

II. GENERAL PRINCIPLES

A. Antibody

Since any staining produced by the fluorescein-labeled antibodies will be interpreted as a reaction with the corresponding antigen, it must be ascertained that the antiserum used reacts only with that antigen; i.e., that it is monospecific. This can be investigated by double-diffusion gel precipitation and immunoelectrophoresis tests. It is advantageous to use antisera with a very high titer of antibodies, and for comparative studies the antibody concentration should be determined.

B. Antigen

Some antigens, such as thyroglobulin, are soluble in the saline solution used to wash the sections; therefore, the tissue has to be fixed to the slide. However, the right fixative must be used since, obviously, the test cannot be performed if the antigen is destroyed during fixation. Generally, tissues are used as soon as possible because some antigens disappear after a short period of time, even though the tissues have been stored at -20°C or even at -70°C [17]. If viable cells are used as a substrate for the test and the antigen is intracellular, then the cell membranes must first be made permeable to the antibody. This can be accomplished by freezing and thawing, by fixation in ether or acetone, etc.

III. IMMUNOFLUORESCENCE TECHNIQUES

A. Optics

The principle of the immunofluorescence test is based on
the observation that when fluorochromes are illuminated with
ultraviolet or blue light they emit light with a longer wavelength. Thus, the color of the light emitted by fluorescein is
green and that emitted by rhodamine is red. Detailed discus-
sions of the physics of fluorescence and the optics that are
essential to view it have been published [11-13]. A high-
pressure mercury lamp is the most frequently used light source
and it is made by many manufacturers [11]. Xenon-arc and
carbon-arc lamps are also available commercially; more recently,
a tungsten-halogen lamp has been introduced [18,19]. The advantages of the tungsten lamp are that maximum illumination is
achieved immediately and that it is cheap. Since it emits little ultraviolet light, it excites little autofluorescence in
tissues but the fluorescence is less intense.

Different filters are available from several manufacturers
[11]. Recently, special types of filters have been developed
that transmit only a narrow band of light (less than 25 nm)
[19]. For fluorescence microscopy, filters with a main trans-
mission band close to the excitation peak of the fluorochrome
are chosen. When such interference filters are used as excitation filters, that is, between the light source and the conden-

ser, the autofluorescence of the tissue is considerably
diminished. Since these filters are more efficient than the
ordinary colored-glass filters, the intensity of the specific
fluorescence is increased. A dark-field condenser must be used;
preferably, it should be of a reflecting or cardioid type.
This type of condenser can be made even more efficient by being
equipped with a toric lens. Several types of such condensers
are available commercially (Tiyoda, Tokyo, Japan and American
Optical Co., Buffalo, New York).

A recent improvement in fluorescence microscopy involves
the use of incident light for excitation of the fluorescence
[19,20]. This has been made possible by the development of
dichroic mirrors of high quality. Such mirrors have multiple
interference coatings, which reflect over 90% of light of short
wavelength but transmit most of the light of longer wavelength.
In this system of illumination, this mirror is placed directly
above the objective at an angle of 45°. The exciting light is
reflected by the mirror into the objective, which functions as
a condenser, concentrating the light on a small area of the
specimen. The resulting fluorescent light is then focused by
the microscope as usual, since it can readily pass through the
mirror into the ocular.

There are several advantages to vertical illumination or
epi-illumination:

(1) The oil immersion dark-field condenser with all its
drawbacks is eliminated.

(2) The exciting light is concentrated on a much smaller area of the specimen.

(3) Since the surface of the specimen facing the objective is illuminated, less light is lost by scatter or by absorption in the specimen.

(4) The specimen can be examined concurrently by transmitted light, by phase contrast, or by dark field for purposes of orientation, etc.

(5) Since the full aperture of the objective is used for illumination, it is advantageous to use high-aperture objectives.

Epiillumination is especially advantageous for work with isolated cells, such as platelets or lymphocytes.

It is of interest that some observers may have difficulty with the immunofluorescence technique because of an inability to recognize the green color of the fluorescing conjugate [21].

B. Labeling with Fluorochromes

Several methods for labeling the globulin fraction of antisera with fluorochromes have been published [11-13]. The globulin fraction can be prepared by precipitation with 50% ammonium sulfate, but care must be taken to remove all the ammonium ions by dialysis against 0.15 M NaCl before conjugation. Alternatively, the globulin fraction may be obtained by column chromatography or by elution of the antibodies from appropriate immuno-

absorbents or from tissues to which the antibodies have been
bound. Preservatives should not be used at this stage because
they may interfere with the protein determinations or with the
labeling process itself.

The labeling procedure that I use is as follows: After
the protein concentration of the globulin solution is determined
by the biuret method, it is adjusted to 20 mg per ml with
0.15 M NaCl. The amount of fluorescein isothiocyanate
(Isomer 1, crystalline, chromatographically pure, Baltimore
Biological Lab., Baltimore, Md.) to be used varies from 5 μg
per mg of protein for very light labeling to 30 μg per mg of
protein for heavy labeling of the globulins. The calculated
amount of fluorescein is dissolved in a volume of 0.5 M carbonate buffer, pH 9.5, equal to 15% of the volume of the globulin solution; it is then very slowly added to the globulin
solution with constant agitation. The final protein concentration is adjusted to 10 mg per ml by addition of 0.15 M NaCl.
The reaction is allowed to take place for about 3 hr at room
temperature or for 18 hr at 4°C.

An alternative labeling method that is especially advantageous for labeling small amounts of globulins is the one
described by Clark and Shepard [22]. The globulin solution is
placed into dialysis tubing. A fluorescein solution of 0.1 mg
per ml is prepared in a 0.05 M carbonate buffer, pH 9.5. The
dialysis tubing containing the globulin solution is suspended

in ten times its volume of the fluorescein solution, and the

reaction is allowed to proceed for 18 hr at 4°C.

The unreacted fluorescein can be removed from the globulin

solution by dialysis, but this takes from four to six days.

Alternatively, it can be removed by gel filtration in

Sephadex G-25. Since heavily labeled antibodies acquire a

high negative charge, they will bind to tissues with the pro-

duction of unwanted nonspecific fluorescence [23]. These

heavily labeled antibodies, as well as the free fluorescein,

can be removed from the conjugate by fractionation on a DEAE-

cellulose column [11-13]. The nonspecific staining produced by

heavily labeled antibodies can be minimized by use of antisera

with high titers of antibodies, so that the conjugate may be

diluted considerably. Absorption with liver (acetone) powder

also can be used effectively. This method is very easy to

implement, but evidence has been presented that the resulting

conjugates deteriorate more quickly in storage.

For purposes of standardization, the fluorescein-to-protein

(F/P) ratio should be determined. The protein concentration is

determined with the biuret reagent, but the reading should be

made at 560 nm instead of at 540 nm to eliminate absorption

owing to the fluorescein in the conjugate. The fluorescein

concentration is measured by the method of McKinney et al. [24].

An empirical method for obtaining the F/P ratio by comparison

of the absorption of the conjugate at 493 nm with the absorption

at 283 nm has also been described [25]. Lightly labeled conjugates have an F/P ratio of 3 to 5, and heavily labeled conjugates have an F/P ratio of 8 to 11.

Similar methods are used for labeling globulins with tetramethyl rhodamine isothiocyanate (Isomer R, crystalline, chromatographically pure, Baltimore Biological Lab., Baltimore, Md.) but about twice as much fluorochrome per mg of protein should be used [13].

C. Processing Specimens

The handling and processing of tissues for immunofluorescence have been described in detail [11-13]. The tissue is generally frozen in liquid nitrogen or in a slurry of dry ice and acetone. Some antigens tolerate storage at -20°C to -70°C for more than a year, whereas others deteriorate very quickly [17]. The tissue also may be sectioned immediately and then stored as sections on slides for future study. Freezing and thawing of the stored tissue renders it useless for immunofluorescence studies because of the marked nonspecific staining encountered in such tissues.

Storage problems are simplified and better histological detail is obtained if the tissue is embedded in paraffin and sectioned at room temperature. One such method is to freeze the tissue in liquid nitrogen and then treat it at -20°C or -70°C with acetone or ethanol, after which it is embedded in paraffin [26,27]. Sections are deparaffinized in tetrahydrofuran or in xylene and then used as usual in the immunofluorescence tests. Freeze-dried specimens also may be embedded in paraffin [11-13]. Some antigens can still be detected by immunofluorescence tests after routine formalin fixation and paraffin embedding. Needless to say, one must always be certain that the antigen is not inactivated if these procedures are being used.

D. Staining Procedures

Details of the staining procedures with fluorochrome-labeled antisera have been published [11-13] and will be described here only briefly. The substrate for the test can be a suspension of organisms or cells, monolayers of cell cultures, tissue imprints or sections of tissues, or artificial substrates with incorporated antigen. The specimen can be used unfixed or it can be fixed by air drying, or with such fixatives as acetone, ether, ethanol, methanol, glutaraldehyde, or formaldehyde. Generally, the histological detail is better if some type of fixative is used.

In the direct immunofluorescence test, an appropriate dilution of the conjugated antiserum is layered on the washed section. After incubation at room temperature for 30-60 min, the slides are washed in phosphate-buffered saline, (PBS), pH 7.3, for 30-60 min, and are then mounted with a nonfluorescent mountant such as a 50-90% solution of glycerol in PBS.

Some investigators have used mounting media with a pH as high as 8.5 to increase the intensity of the fluorescence. This test is used primarily to detect antibody already bound to the antigen, and to detect antigens in the specimens.

In the indirect immunofluorescence test, the washed section is overlaid with an appropriate dilution of the test serum. After incubation for 30-60 min at room temperature, the section is washed for 30-60 min in PBS; the labeled antiserum is then applied as in the direct test. This test is used primarily to detect anibodies present in sera and eluates.

Complement fixation also can be performed on the sections. The washed slide is treated with a heat-inactivated test serum. After 30-60 min, the slide is washed; then, fresh serum from another species or serum from another animal of the same species is layered on the section. After incubation for 30-60 min, the section is treated with a fluorescein-labeled antiserum to the complement of the species used in the test.

The slides are generally washed in Coplin jars. A different method that we have found useful is to put the slides into a staining rack and then to suspend the rack in a staining dish. The PBS is agitated with a magnetic stirrer. After the slides are mounted, it is advantageous to fix the coverslips in place by painting the edges with nail polish. This procedure prevents the sections from being destroyed by inadvertent movement of the coverslips.

One of the drawbacks of the immunofluorescence technique
is that the stain fades quickly; as much as 30% may be lost
overnight. A method of preserving the fluorescence for at least
two to four months involves drying of the stained slide for
1 min, fixing the section in 95% alcohol for 15 to 30 min, drying, and mounting as usual [28].

E. Tests for Specificity

Final tests for the specificity of the fluorescein-labeled
immunoglobulin are performed with tissue sections or other convenient substrate. Such tests include inhibition of the staining reaction by layering first the unlabeled antiserum on the
substrate, followed by the labeled antiserum, and by absorption
of the labeled antiserum with the purified antigen. Better
controls of specificity are the internal ones; for example, if
a tissue section is used, only the specific structure in ques-
tion should be stained. If the purpose of the test is to stain
for in vivo-bound immunoglobulins in immune complex nephritis,
for example, further assurance of specificity is obtained if
normal glomeruli do not stain, some diseased glomeruli are
stained only by an anti-IgG, others only by an anti-IgA, and
still others only by an anti-IgM fluorescein-labeled antiserum
[29]. The specificity of antiimmunoglobulin conjugates can be
tested also by the direct immunofluorescence test on samples of
bone marrow obtained from patients with IgG or IgA multiple

myeloma or with Waldenström's macroglobulinemia [30]. Alternatively, the test antigen can be incorporated into an artificial substrate such as normal rabbit serum cross-linked with glutaraldehyde that is then sectioned and stained as usual [31]. The optimal titer of the conjugate is also determined by a test of serial dilutions on positive sections and use of the highest dilution that still gives strongly positive staining.

To determine whether the detected immunoglobulins in diseased glomeruli, for example, are specifically bound, as opposed to being trapped nonspecifically, it should be ascertained whether or not the complement is also bound in a similar pattern. Further corroboration of the antibody nature of the detected immunoglobulin can be obtained by elution of the immunoglobulin from the diseased glomeruli and demonstration of its antibody activity in vitro. For example, one may perform virus neutralization tests, precipitation, etc. An ingenious way of doing this on tissue sections has been developed [32]. For this procedure, 0.2 ml of 0.02 M citrate buffer, pH 3.2, was incubated overnight on the tissue section at 4°C. Then the drop was aspirated into a calibrated syringe and neutralized with a calculated amount of 1 M NaOH. This solution containing the eluted antibody was then placed on sections of other tissues, and the presence of the antibodies was demonstrated by immunofluorescence.

To inhibit the nonspecific fluorescence that may be a prob-

lem with some specimens, counterstaining can be used. Simple
stains that provide a red fluorescence against which the specific green fluorescence of the labeled antibody is observed
are 0.01-0.005% solutions of Congo red and Evans blue. Papain
or serum albumin labeled with rhodamine also can be used for
this purpose.

Special problems can be encountered in attempts to demonstrate an antigen in immune complexes. In some cases the
antigens appear to be completely covered by immunoglobulins,
which then must first be partially eluted before the hidden
antigen can be detected. This can be performed on tissue sections by incubation of the slides for 30 min in 0.02 M citrate
buffer, pH 3.2. In some cases elution with 2 M NaCl solution
[33] or with 2.5 M potassium thiocyanate solution [34] appears
to be more effective.

It is possible to detect two antigens simultaneously by
use of a fluorescein-labeled antiserum to one antigen and a
rhodamine-labeled antiserum to the other antigen. The conjugates may be applied sequentially or simultaneously. The
fluorescein-labeled antibody will fluoresce green and the
rhodamine-labeled antibody will fluoresce red; if the antigens
are in close proximity to each other, the combined fluorescence
will have an orange color. Appropriate filters should be used
for such tests [35].

F. Photomicrography

For color photomicrography Ektachrome daylight film,
ASA 160, and for black and white prints Kodak Tri-X Pan film,
ASA 300, give good results. Exposure times vary from 10 to 90
sec. Exposure meters are available, but are not used routinely.
However, microphotometry has been used to quantitate the fluorescence in both the direct and indirect antibody techniques
[36,37].

G. Applications of the Immunofluorescence Technique

The immunofluorescence technique is finding increasingly
wide application in many fields. The direct test, in which
flurochrome-labeled antibody is used, has been used to detect
antibodies and complement bound in vivo to tissue antigens and
to determine the location of various antigens in various cells
and tissues. Tests with labeled antigen also have been performed. The indirect test has been used to detect serum antibodies to various tissue, viral, bacterial, fungal, and para-
sitic antigens. It has also been used to detect soluble antigens
that have been made insoluble by polymerization with glutaraldehyde or that have been incorporated into various insoluble
substrates. Soluble antigen also has been coated on agarose
beads, and then demonstrated by immunofluorescence tests [38].
Further, equipment has been designed to perform automated indirect immunofluorescence tests on various antigens [39]. Lasers

have been designed to excite the fluorescence, which is up to a

thousand times as intense as that excited by mercury lamps [40].

IV. IMMUNOENZYME TECHNIQUE

This technique is based on the principle that an enzyme

can react with many molecules of its substrate to produce reac-

tion products that are insoluble in organic solvents and,

depending on the substrate used, of different colors. Since

some of the reaction products are also electron-dense and

stable under an electron beam, enzymes can also be used as

labels for antibodies in electron microscopy studies. While

different enzymes can be used, those most frequently used are

horseradish peroxidase, phosphatases, and cytochrome C [6].

Some investigators have used high-titer antisera, whereas

others have found it necessary to prepare purified antibodies

with the aid of immunoabsorbents. Labeling has been performed

in different ways for different enzymes. The simplest and most

satisfactory way, however, appears to be the use of glutaralde-

hyde to cross-link the enzyme to the antibody [6]. The use of

the purest enzyme preparations available is recommended. The

conjugation procedure is as follows: 12 mg of peroxidase is

added to one ml of 0.1 M phosphate buffer, pH 6.8, containing

5 mg of purified antibody. The solution is stirred gently and

0.05 ml of a 1% aqueous solution of glutaraldehyde is added

dropwise. The mixture is allowed to stand for 2 hr at room temperature, and is then dialyzed for 18 hr against two changes of 5 liters of buffered saline at 4°C. The precipitate formed is removed by centrifugation at 20,000 x g for 30 min. The enzyme-labeled antibody preparations stored at 4°C showed no appreciable loss of catalytic or immunologic specificity for at least several months.

For light-microscopy studies with the immunoenzyme technique, the procedures used in the immunofluorescence technique are used. However, an additional step must be performed to demonstrate the enzyme used as a label. Different substrates, which yield different reaction products, must be used for the different enzymes. For horseradish peroxidase, the sections are incubated in a saturated solution (75 mg/100 ml) of 3,3-diaminobenzidine (free base) in 0.05 M Tris·HCl buffer, pH 7.6, and 0.001% H_2O_2 for 30-60 min. The sections are then washed, dehydrated, cleared, and mounted.

A method for the localization of antigens in tissues without conjugation of the corresponding antibody with the enzyme also has been developed [41]. This procedure involves the sequential reaction of the following reagents: (1) specific antiserum to the tissue antigen, (2) antiserum to the immunoglobulin of the species used for preparation of the antibody in (1), (3) specific antiserum to the enzyme to be used, prepared in the same species as in (1), (4) the enzyme itself, and (5)

the substrate solution for visualizing the enzyme. This bridge
method has the advantage that many different antibodies can be
detected with one indicator system, and the antibodies do not
have to be conjugated.

A second indirect immunoenzyme technique involves the
preparation of a hybrid antibody that will react with one antigen at one combining site and with a second antigen at the other
combining site. Thus, the one antigen could be an immunoglobulin, for example, and the second antigen could be a peroxidase
[6]. To demonstrate the localization of the first antigen, the
tissue is first incubated with the hybrid antibody, then with
peroxidase, and finally the specimen is stained for peroxidase.

Another useful modification of the immunoenzyme technique
is the following. After reaction of the section with the
peroxidase-labeled antibody, peroxidase is added. Then an
antiperoxidase antibody is added, and finally more peroxidase
is added. The specimen is then stained for peroxidase. This
modification has the effect of enhancing the sensitivity of the
method.

Procedures for the ultrastructural location of enzyme-
labeled antibodies are more complicated. Since the penetration
of the labeled antibodies into a tissue specimen is limited,
only light fixation must be used [6]. Several different fixatives can be used. The test is performed on 0.5- to 1-mm
blocks of tissue, on 10- to 50-μm frozen sections obtained with

a cryostat and on ultrathin frozen sections [6]. It can be
performed also on ultrathin sections of methacrylate-embedded
tissues [42]. Peroxidase is revealed by the diaminobenzidine
and hydrogen peroxide methods and phosphatase is revealed with
Gomori's medium [6].

It is possible to detect simultaneously at least two different antigens in tissue specimens by labeling the two antisera
with different enzymes or by using different substrates to produce reaction products of different colors if both antisera are
labeled with the same enzyme [6,43]. Other ways of doing this
are to label one antibody with fluorescein or ferritin and the
other with peroxidase [6].

Although the immunoenzyme technique has not been studied
in as great detail as the immunofluorescence technique, it
appears to have similar applicability to the study of immunology
and immunopathology. For the immunofluorescence technique, a
special optical system and light source, as well as a dark-field
condenser, must be used. The slides must be examined in a
darkroom. The immunofluorescence fades quickly; as much as 30%
may disappear overnight and, under illumination, the fluorescence also fades very quickly. For photographic purposes,
exposure times of 10 to 90 sec are needed. For the immunoperoxidase technique an ordinary light microscope is all that
is necessary. The test itself is somewhat longer, since the
section has to be treated with the appropriate substrate to

develop the reaction products. However, the slides keep well

for at least one to two years. Conjugation of antibodies with

peroxidase also is easy if glutaraldehyde is used to cross-link

the enzyme to the antibody. The two methods have about equal

sensitivity in detecting antinuclear antibodies [44], tissue

antibodies [45], and bound immunoglobulins in human renal biop-

sies [46]. With either method at least two antigens can be

detected simultaneously in tissue sections or other substrates.

However, peroxidase can be used also as an in vivo tracer,

whereas fluorescein-labeled compounds are quickly removed by

the liver. Further, only the immunoperoxidase technique can

be used for ultrastructural studies.

The ultrastructural localization achieved with the immuno-

peroxidase technique is as good as that obtained with ferritin

[6]. In some respects, it even appears to be superior to the

immunoferritin technique. For example, better localization of

cell-surface antigens could be achieved with it [47].

Peroxidase-labeled antibodies exhibit better tissue penetration,

probably because of the size of the two markers; peroxidase has

a molecular weight of 40,000, while ferritin has a molecular

weight of 460,000. Peroxidase-labeled antibodies can be used

also as in vivo tracers, whereas ferritin-labeled antibodies

localize only poorly [48]. The immunoenzyme method is also

more sensitive than the immunoferritin method, and the back-

ground staining is considerably less [6].

REFERENCES

1. A. G. E. Pearse, Histochemistry. Theoretical and Applied, Vol. I, 3rd ed., Churchill, London, 1968, 759 pp.

2. G. M. Barenboim, A. N. Domanskii, and K. K. Turoverov, Luminescence of Biopolymers and Cells (Translated by R. F. Chen), Plenum Press, New York, 1969, 229 pp.

3. G. L. Wied and G. F. Bahr (eds.), Introduction to Quantitative Cytochemistry II, Academic Press, New York, 1970, 551 pp.

4. H. J. Barr and J. R. Ellison, Nature, 233, 190-191 (1971).

5. W. Bodmer, M. Tripp, and J. Bodmer, Histocompatibility Testing (E. S. Curtoni, G. Mattinz, and R. M. Tosi, eds.), Williams and Wilkins, Baltimore, Md., 1967, pp. 341-350.

6. S. Avrameas, Intern. Rev. Cytol., 27, 349-385 (1970).

7. T. K. Shnitka and A. M. Seligman, Ann. Rev. Biochem., 40, 375-396 (1971).

8. M. G. Farquhar, S. L. Wissig, and G. E. Palade, J. Exptl. Med., 113, 47-66 (1961).

9. R. C. Graham, Jr., and M. J. Karnovsky, J. Exptl. Med., 124, 1123-1134 (1966).

10. G. C. Budd, Intern. Rev. Cytol, 31, 21-57 (1971).

11. M. Goldman, Fluorescent Antibody Methods, Academic Press, New York, 1968, 303 pp.

12. R. C. Nairn, Fluorescent Protein Tracing, 3rd. ed., Williams and Wilkins, Baltimore, Md., 1969, 502 pp.

13. A. Kawamura, Fluorescent Antibody Techniques and Their Application, University Park Press, Baltimore, Md., 1969, 203 pp.

14. G. A. Andres, K. C. Hsu, and B. C. Seegal, Handbook of Experimental Immunology (D. M. Weir, ed.), F.A. Davis Co., Philadelphia, Pa., 1967, pp. 527-570.

15. S. S. Breese, Jr. and K. C. Hsu, Chapter 3 in this volume.

16. M. A. Benjaminson, Stain Technology, 44, 27-31 (1969).

17. E. Barnett, Standardization in Immunofluorescence (E. J. Holborow, ed.), Blackwell Scientific Publications, Oxford, England, 1970, pp. 75-84.

18. G. V. Heimer and C. E. D. Taylor, J. Clin. Pathol., 25, 88-93 (1972).

19. J. S. Ploem, Ann. N. Y. Acad. Sci., 177, 414-429 (1971).

20. S. S. Kasatiya and A. Birry, Am. J. Clin. Pathol., 57, 395-399 (1972).

21. G. Sander, Standardization in Immunofluorescence (E. J. Holborow, ed.), Blackwell Scientific Publications, Oxford, England, 1970, pp. 155-158.

22. H. F. Clark and C. C. Shepard, Virology, 20, 642-644 (1963).

23. D. Bruchhausen, R. Hofermann, and H. V. Mayersbach, Immunology, 19, 1-10 (1970).

24. R. M. McKinney, J. T. Spillane, and G. W. Pearce, Anal. Biochem., 14, 421-429 (1966).

25. H. P. Brusman, Anal. Biochem., 44, 606-611 (1971).

26. G. Sainte-Marie, J. Histochem. Cytochem., 10, 250-256 (1962).

27. R. S. Post, Cryobiology, 1, 261-269 (1965).

28. J. Bienenstock and J. Dolezel, J. Histochem. Cytochem., 18, 518 (1970).

29. J. Klassen, R. T. McCluskey, and F. Milgrom, Am. J. Pathol., 63, 333-358 (1971).

30. W. Hijmans, H. R. E. Schuit, A. P. M. Jongsma, and J. S. Ploem, Standardization in Immunofluorescence (E. J. Holborow, ed.), Blackwell Scientific Publications, Oxford, England, 1970, pp. 193-202.

31. H. Brandzaeg, Immunology, 22, 177-183 (1972).

32. T. E. W. Feltkamp and J. H. Boode, J. Clin. Pathol., 23, 629-631 (1970).

33. D. Koffler, P. H. Schur, and H. G. Kunkel, J. Exptl. Med., 126, 607-624 (1967).

34. T. S. Edgington, R. J. Glassock, and F. J. Dixon, Science, 155, 1432-1434 (1967).

35. G. D. Johnson, Standardization in Immunofluorescence (E. J. Holborow, ed.), Blackwell Scientific Publications, Oxford, England, 1970, pp. 123-125.

36. W. J. Irvine, M. M. W. Chan, and W. G. Williamson, Clin. Exptl. Immunol., 4, 607-617 (1969).

37. D. Gandini-Attardi, J. B. Fleischman, O. S. Pettengill, and G. D. Sorenson, Cellular Immunol., 2, 101-113 (1971).

38. A. Bürgin-Wolff, R. Hernandez, and M. Just, Experientia, 28, 119-120 (1972).

39. A. Birry, M. Caloenescu, and S. S. Kasatiya, Am. J. Clin. Pathol., 57, 391-394 (1971).

40. G. I. Kaufman, J. F. Nester, and D. E. Wasserman, J. Histochem. Cytochem., 19, 469-476 (1971).

41. T. E. Mason, R. F. Phifer, S. S. Spicer, R. A. Swallow, and R. B. Dreskin, J. Histochem. Cytochem., 17, 563-569 (1969).

42. Y. Kawarai and P. K. Nakane, J. Histochem. Cytochem., 18, 161-166 (1970).

43. W. Strauss, J. Histochem. Cytochem., 20, 272-278 (1972).

44. J. Dorling, G. D. Johnson, J. A. Webb, and M. E. Smith, J. Clin. Pathol., 24, 501-505 (1971).

45. V. Petts and I. M. Roitt, Clin. Exptl. Immunol., 9, 407-418 (1971).

46. F. R. Davey and G. J. Busch, Am. J. Clin. Pathol., 53, 531-536 (1970).

47. R. Bretton, T. Terynyck, and S. Avrameas, Exptl. Cell. Res., 71, 145-155 (1972).

48. P. Druet, J. Bariety, B. Bellon, and F. Laliberte, Lab. Invest., 27, 157-164 (1972).

Chapter 5

CHARACTERIZATION OF ANTIGENS FROM ONCOGENIC VIRUSES

Theodore P. Zacharia

National Academy of Sciences
Washington, D. C.

I.   INTRODUCTION 104

II.   WHAT INFLUENCES THE YIELD OF ANTIGEN? 105

      A. Best Antisera 106

III. SOURCE OF ANTIGENS 106

IV.   CELLS TRANSFORMED BY RNA ONCOGENIC VIRUSES 109

      A. Cell Antigens 109
      B. Viral Antigens 118

V.   CELLS TRANSFORMED BY DNA ONCOGENIC VIRUSES 128

      A. Complement-Fixing and Transplantation
          Antigens 128
      B. Group-Specific Antigens 133

      REFERENCES 134

I. INTRODUCTION

The most important step in an investigation of the immune response to oncogenic viruses is the isolation and purification of antigens and antibodies. This can present several problems, the most serious of which is the alteration of antibody and antigen molecules during isolation and purification. It is advisable to check the various properties of antigens and antibodies after isolation and purification.

Concentrations of antigens and antibodies vary depending on the source and the methods of isolation. The tissue that gives the highest yield in the purest form attainable is the most appropriate source. This is especially important in the case of neoplasia induced by oncogenic viruses. An examination of the tumor and surrounding tissue is very important.

A difficulty encountered by several investigators is the possible inaccessibility of antigens and antibodies, which can be masked by compartmentation, deposits, complex formation, lack of virus helper, and/or absence of symptoms.

Different hosts respond to an immunogen in different ways and to different extents. Young animals, especially at a very early age, are immunologically paralyzed by smaller doses, in general, than would be necessary for older ones. The induction of immune tolerance or paralysis depends also on the route of inoculation of the antigen and on the solvent used.

II. WHAT INFLUENCES THE YIELD OF ANTIGEN?

The yield of antigen was found to be affected by several factors.

(1) Route of infection. Cells infected in suspension gave higher antigen titer than did plated cells.

(2) Maintenance medium. Calf serum in the medium gave better results (4-32 times higher antigen titers) than agamma-globulinemic calf or horse serum.

(3) Time and method of harvest. With optimal doses of inoculum [e.g., Rauscher leukemia virus, 10^5 focus-forming units (FFU)], optimal harvest time is 6-14 days after infection. With lower doses of virus (100 FFU), highest antigen titers were attained 18-22 days after infection.

Higher antigen titers were associated with cells, not with the culture medium. Antigen was obtained by sonication or grinding of infected cells.

It is important to know whether the virus persists in tumor cells. It does in infections with avian leukosis, murine leukemia, feline leukosis, and poxviruses. However, the virus does not persist in tumor cells infected with adeno-, herpes-, polyoma-, Simian- (SV 40), and papillomaviruses (the last three form the papova group of viruses).

A. Best Antisera

The best antisera are those with the highest titer when
tested against a specific antigen. For example, in the test for
detection of the murine leukemia virus, antisera were obtained
from rats with a massive subcutaneous tumor that metastesized
to the local immunocompetent organs (lymph nodes); these rats
survived a long period after inoculation with oncogenic virus
or with transplants from a tumor induced by the Gross leukemia
virus.

III. SOURCE OF ANTIGENS

There are two main sources of antigens in animals infected
with RNA and/or DNA oncogenic viruses (see Tables 1 and 2),
antigens of cellular origin and antigens of viral origin.

Cells transformed by RNA or DNA oncogenic viruses possess
surface (transplantation) antigens [42] and internal (humoral
or serological) antigens. RNA and some DNA oncogenic viruses
possess virion group-specific antigens (see below), enzyme
antigens (DNA-directed DNA polymerase in adeno- type 12, poly-
oma-, simian- 40, papilloma-, human wart, and herpesviruses;
RNA-directed RNA polymerase in leukemia, sarcoma, and human-
milk viruses; RNA-directed DNA polymerase (reverse transcrip-
tase [55]), and capsid antigens.

TABLE 1

Major Oncogenic Viruses[a]

Virus	Host
DNA viruses	
Papova	
Papilloma	Man (wart), cow, dog
Polyoma	Mouse
Simian 40	Monkey
Adeno	Man, cow, monkey, bird
Herpes	Man, birds
Burkitt's lymphoma	Man
Lucké's carcinoma	Frog
Marek's disease	Bird
RNA viruses	
Leukemia-sarcoma	Cat, rat, guinea pig, mouse, hamster, bird
Mammary tumor	Mouse
Milk	Human
RD (rhabdosarcoma)114	Human

[a]See Table 2 for references.

TABLE 2

Isolation and Purification of Antigens

Source of antigen (virus)	Reference(s)
Adeno	43
Avian myeloblastosis	7, 8, 12, 27, 62
Epstein-Barr	95
Feline leukemia	31, 35, 57, 77, 87, 88
Hamster leukemia	35, 96, 97A
Lucké's carcinoma	82
Mammary tumor	45
Marek's disease	1, 109
Milk (human)	55
Murine leukemia	30, 35, 41, 44, 45, 61B, 87, 88, 105
Murine sarcoma	35, 44, 57
Papilloma	71, 92
Polyoma	21, 97B, 99, 100
RD (rhabdosarcoma)114	77
Rous sarcoma	8, 12, 25, 27, 71, 84
Simian 40 (SV40)	13, 20, 21, 100
Wart (human)	71

IV. CELLS TRANSFORMED BY RNA ONCOGENIC VIRUSES

A. Cell Antigens

Transplantation antigens can be detected by (1) fluorescence antibody tests and (2) cytotoxicity tests. Antigens that
occur inside a cell include various factors formed during virus
replication, virus-cell interaction, and cell transformation
[Figs. 1 and 3 in Ref. 39], and serological antigens that occur
in the cell cytoplasm or nucleus, and can be detected by (3)
fluorescein-antibody tests, (4) cytotoxicity tests, (5) complement fixation tests, (6) hemagglutination, (7) hemagglutination
inhibition, and (8) indirect hemagglutination.

1. Immunofluorescence

The immunofluorescence tests are performed with sera from
tumor-bearing animals and autologous or syngeneic cells.

a. Indirect Immunofluorescence. Surfaces of tumor cells
stain in the presence of a specific antiserum, but do not stain
in the presence of nonspecific sera or sera from uninfected
animals. The following test procedure was used.

(i) Slides were used that were cleaned with alcohol and
distilled water, and dried (this cleaning is very important).

(ii) Moistened pieces of filter paper were evenly spaced
on the slide.

(iii) The slide was sprayed with Fluoroglide (Fisher) and allowed to dry.

(iv) Pieces of filter paper were removed, leaving wells surrounded by hydrophobic films.

(v) A cell suspension was prepared with a cell density of about 10^6 cells/ml.

(vi) One drop of the cell suspension was added to each well.

(vii) Cells were dried for 1-2 hr at 37°, fixed with acetone, and stored at -20° to -70°. Tissue culture cells cultivated in the wells of the slides showed more detail by immunofluorescence than did trypsinized tumor cells that were allowed to dry on the slide, and which show rounded nuclei and contracted cytoplasm.

(viii) A drop of antiserum dilution was added to the wells. The cell-antiserum mixture was incubated for 1 hr at 37° in a humidified incubator; slides were washed with phosphate-buffered saline (PBS), pH 7.4 (twice), and distilled water (twice). A drop of fluorescein-labeled antiglobulin was added to each well; slides were then incubated at 37° for 1 hr and washed two or three times with PBS and twice with distilled water. Then 0.06% of Evan's blue stain was added (counterstain) and the mixture was kept 4-5 min. Stained cells were washed with distilled water and dried. Cells were mounted in glycerine-PBS and stored covered at -20°. Fluorescence lasted almost a year.

(ix) Cells were examined under a microscope with a dark-
field condenser.

b. Immunofluorescence Adsorption Test. The immunofluorescence adsorption test is a test for viral antigens.

Antisera to viruses were adsorbed with cell extracts, and
adsorbed sera were tested by immunofluorescence for residual
antibodies specific to the tested virus. Powders of lyophilized
tissue or tissue homogenates were used for adsorption. Tissues
were disrupted by sonication in isotonic saline or Veronal buffer (pH 7.4). Fragmented tissues were homogenized in 25-30 vol
of distilled water and centrifuged at 48,000 x g for 2.5 hr.
The soluble fraction was lyophilized.

Sera diluted two or three double dilutions beyond the end
point of fluorescence were refrigerated at 4° for 18 hr with the
lyophilized tissue powder or homogenate (60 mg of the tissue
preparation per millileter of diluted serum or antiserum). The
serum-tissue mixture was centrifuged for 30 min at 1000 x g.
The antigen titer was the maximum dilution of tissue preparation
that adsorbed specific fluorescence from the diluted antiserum.

2. Cytotoxic Test

Antibody against cellular antigen reacts with the cell in
the presence of complement. This reaction causes death of the
cell, which can be detected by (i) failure of dead cells to

grow in the isologous host, (ii) staining with the dye trypan
blue, and (iii) release of ^{51}Cr from damaged cells previously
labeled with the isotope.

  a. Microcytotoxicity Test. The 200-400 fibroblasts or
tumor cells dispersed by trypsinization (0.03% trypsin) were
seeded in 0.25 ml of minimal essential medium in each well of a
test plate and incubated at 37° for 18 hr to allow cell adherence. Plating efficiency ranged between 40% and 75%. Adherent
target cells were washed with the medium; and 0.5 ml of 1:4
diluted, heat-inactivated (56° for 30 min) serum (agammaglobu-
linic fetal-calf serum) was added to each well. Cells were
incubated at 37° for 45 min. Then 0.05 ml of medium, 2-4 x 10^5
test lymphocytes, or normal lymphocytes were added to each well.
Incubation was continued for 60 min at 37°, when 0.15 ml of
growth medium with fetal-bovine serum was added to each well,
and incubation was continued for 60 hr at 37°. Cells were fixed
with methanol and stained with Giemsa. The total number of target cells per well was determined.

  Tumor and normal target cells were cultured in Eagle's
minimal essential medium with 2 x vitamins, 2 x glutamine, nonessential amino acids, biotin (1 mg per liter), 5% NCTC 135,
75 μg of streptomycin/ml, 120 units of penicillin/ml, and 20%
of heat-inactivated fetal-bovine serum. Lymphocytes were isolated from heparinized peripheral blood sedimented with plasma
gel (Laboratoire Roger Bellon, Neuilly, France; 3:1, v/v) for

30 min at 24°. Plasma with leukocytes was then incubated on a washed nylon column at 37° for 45 min. Columns were eluted with minimal essential medium, and the eluate was centrifuged at 100 x g for 10 min at 24°. The pellet was treated with 20 mM Tris·HCl-50 mM ammonium chloride, pH 7.8, which causes erythrocytes to lyse. The lymphocytes were then washed with minimal essential medium, counted, and suspended in the same medium with 10% heat-inactivated fetal-bovine serum.

b. In Vitro Cytotoxic Test. Mice were infected with 5 x 10^3 leukemia-producing units/mouse. A leukemia-producing unit is the highest dilution of virus that produced hepato-splenomegaly six weeks after inoculation of DBA/2 mice.

Spleens were obtained from mice 4-5 weeks after infection. Spleen cells from leukemic and normal mice were labeled with ^{51}Cr by incubation of 10^7 cells with 180 μCi of ^{51}Cr (as sodium chromate). Labeled cells were washed three times with PBS for 30 min at 4°, and suspended in minimal essential medium at a density of 4 x 10^6 cells/ml. Cells (2 x 10^6), 0.6 ml of 10% mouse serum, and 3.3% guinea-pig complement were incubated at 37°. Aliquots were removed at various intervals, and ^{51}Cr in the cell-free supernatant was measured in a well-type gamma scintillation counter.

c. Assay of Cytolysis by Measurement of ^{51}Cr Released. Target monolayers were prepared as described in the preceding test. Then 4 x 10^6 viable lymphocytes were plated on monolayers

of target cells, and ^{51}Cr that was released after incubation at
37° for 24-48 hr was measured in a well-type scintillation counter with a NaI crystal.

3. Virus Neutralization

a. In Vivo Neutralization. Dilutions of test virus or
antigen were mixed with an equal volume of normal or immune
serum and incubated for 30 min at 37°. Four- to six-day-old
mice were inoculated intraperitoneally. Mice were checked
daily for palpable tumors and other symptoms, e.g., splenomegaly,
hepatomegaly, etc.

b. In Vitro Neutralization. Plaque or focus titers were
determined in tissue culture. Dilutions of heat-inactivated or
unheated serum (56°, 30 min) were incubated with equal amounts
of the specific virus for 30 min at 37°. As controls, cells
were incubated with virus alone, virus plus normal serum, and
normal serum alone. Foci or plaques were counted, usually
5-7 days after infection. The following formula resulted:

$$\%(FPFU) = 1 - \left(\frac{(FPFU/ml)_{cv} + (FPFU/ml)_{cs}}{(FPFU/ml)_{cv}} \right) \times 100$$

where FPFU represents focus- or plaque-forming units; CV indicates, from culture with virus alone; and CS indicates, from
culture with immune or normal serum.

c. Leukemia-Virus Neutralization. Dilution of sera

(normal or immune)-500 leukemia-producing units of virus were

incubated at 4° for 1 hr. The mixture was inoculated into mice.

Leukemia was determined by development of splenomegaly (spleens

weighing more than 250 mg). The antibody titer is the highest

dilution that inhibits the development of splenomegaly.

4. Other Immunological Procedures

   Table 3 lists other immunological procedures and the cor-

responding references.

TABLE 3

Other Immunological Procedures

Procedure	Reference(s)
Adoptive transfer	10, 24
Antiserum-complement lysis	91
Radioautography	20
Cell fractionation	101
Cell fusion	20
Cellular immunoadsorbance	103
Colony-forming assay	29, 91
Complement fixation	41, 43, 58, 61B, 79, 88, 89, 106
Cytolysis	91, 103
Cytophilic antibody	15

(Table 3 continued)

TABLE 3 (continued)

Procedure	Reference(s)
Cytotoxicity	26, 48, 51, 62, 70, 80
Delayed hypersensitivity	15, 22, 33, 42
Derivatized fibers and beads	85
Double antibody	88
Early interference tests	7, 94
Electron microscopy	16, 61B
Fiber fractionation of cells	85
Gel diffusion (immunodiffusion)[a]	8, 35
Genetic analysis	61B
Hemagglutination	15, 105
Hemagglutination inhibition	21
Immunoadsorbance	49
Immunoelectron microscopy	16, 95
Immunoelectrophoresis	8, 62
Immunofluorescence	21, 26, 47, 99, 104
Iodination	87, 88
Isoelectric focusing	36, 87, 88
Isotopic antiglobulin	99, 100A, 105
Karyotype determination	95

[a]A mixture of 1% Difco agar-8% NaCl is used with chicken sera, and 1% agar-0.5% NaCl with rabbit or hamster sera. Incubate antigen-antibody reaction two to five days in a moist chamber at 20°.

TABLE 3 (continued)

Procedure	Reference(s)
Leukemia-producing units	105
Linkage testing	61B
Lymphocyte culture	67
Lymphocyte-mediated cytolysis	103
Macrophage-inhibition test	22, 33
Microcomplement fixation	58
Microcytotoxicity	59
Microhemagglutination	109
Neutralization	30, 41, 80
Pathology	61B
Periodic acid-Schiff (histochemical)	12, 61B, 110
Plaque assay	50, 61B, 85
Polyacrylamide gel electrophoresis	12, 30, 34, 87, 90, 102, 108
Purification of group-specific antigen	87
Radioimmunoassay	37, 88, 107
Radioimmunoprecipitation	87, 88
Rosette formation	40, 50, 85
Solid-phase immunoadsorbance	81
Transplantation	48, 53
TD 50 determination	99
Viral expression-host marker	61B
Virus neutralization	60

B. Viral Antigens

Viral antigens include group-specific (gs) (Table 4) and
envelope antigens and other viral proteins, e.g., enzymes
[14,39], and can be detected by all tests mentioned for cell
antigens.

1. Group-Specific Antigens

Group-specific antigens are associated with the virion and
are distinct from the viral envelope protein. They are species
specific, soluble, ether resistant, and stable. The gs antigens
characterize C-type viruses, regardless of the cells used to
grow the virus. This antigen can be used as a marker for the
detection of viral gene activity under conditions where infectious virus, or even viral particles, could not be detected.
The gs antigens contributed to the theory that viral genetic
information is inherited as part of the cellular genome.

The gs antigens of leukemia virus (C-type RNA tumor virus)
were detected in most animal embryos tested, especially mouse
embryos. Younger embryos were more likely to be positive than
older ones, particularly in strains of animals that were negative postnatally with respect to the RNA genome of oncogenic
virus. The gs antigens were detected in embryos of strains
with a low incidence of leukemia that were free of infectious
virus. Thus, the genome of RNA tumor viruses is switched off

TABLE 4

Methods Used to Characterize Group-Specific (gs) Antigen

Method	Reference(s)
1. Polyacrylamide gel electrophoresis	3, 30, 78, 90, 102
2. Determination of N-terminal amino-acid residues with ^{14}C-fluorodinitrobenzene, hydrolysis, thin-layer chromatography on silica, and autoradiography	2, 68, 78
End-group determination	
Amino-terminal sequence of gs-a	68
Pro-Val-Val-Ile-Lys-Thr-Glu-Gly-Pro-Ala-Trp-Thr-Pro-Leu-Glu-Pro-Lys-Leu-Ile-Thr-Arg-Leu-Ala	
Amino-terminal sequence of gs-b	68
Asp-Ala-Met-Thr-Met-$^{Glu}_{Thr}$-His-Lys-Asp-Arg-Pro-Leu-Val-Arg-Val-Ile-Leu-Thr-Ser-Thr-Gly-Ser-His-Pro-Val-	
3. Determination of C-terminal amino-acid residues	2
4. Determination of molecular weights gs-a 20,000, gs-b 11,000	3
5. COFAL (complement fixation avian leukosis)	86
6. Microcomplement fixation	43, 58, 89
7. Immunoprecipitation	31, 32, 35
8. Isoelectric-point determination: chicken (8.9); mouse (6.7); rat (8.6); hamster (6.9); cat (8.3)	98, 74, 36, 75, 76
9. Neutralization	41

and is inherited as part of the genetic apparatus of normal
cells.

    a.   Host Range of Group-Specific Antigens of Leukemia
Viruses. Antigens 1 and 2 (intraspecific antigen; 35) are present in murine leukemia virus, but are absent from feline leukemia
virus, hamster D-9 lymphosarcoma, rat Novikoff hepatoma, and
chicken BAI strain A of avian myeloblastosis virus.

    Antigen 3 (interspecific antigen; 35) is shared by mouse,
cat, hamster, and rat leukemia viruses [32].

    b.   Rat and Hamster Leukemia Viruses. Rat leukemia virus
was obtained from the peritoneal fluid of rats bearing Novikoff
ascites hepatoma [96]. Hamster leukemia virus was obtained
from D-9 lymphosarcoma [97A]. The virus was released from the
tumor by treatment with ether. Viruses were isolated by
density-gradient centrifugation in potassium tartrate [32].
The same technique can be used for isolation of murine and
feline leukemia viruses. The D-9 lymphosarcoma virus of hamsters
banded at 1.3 g/cm^3 [73]; Novikoff rat virus banded at 1.15 g/cm^3
[32]. [For further information on the isolation and characterization of gs-1, see Refs. 32 and 61B; for information on gs-3 of
feline leukemia virus, see Refs. 32 and 87.]

    For detection of gs antigen, the Ouchterlony immunodiffu-
sion test was used with 2% agar in PBS, pH 7.0. The antiserum
against murine leukemia virus was placed in the center well.
Antigens were disrupted virions or extracts of infected tissue [32].

c. <u>Mouse Leukemia Virus Grown in Tissue Culture.</u> Cells
from Swiss NIH mouse embryos were most sensitive to leukemia
viruses with respect to the production of gs complement-fixing
antigen.

Age 17- to 20-day mouse embryos were minced and trypsinized.
Then 8-12 x 10^3 cells/ml were cultured in a growth medium, consisting of 10% unheated fetal-calf serum in Eagle's minimal
essential medium (MEM). All media contained 2 mM glutamine,
100 units of penicillin, and 10 µg of streptomycin per milliliter.

Primary monolayer cultures were treated with 0.25% trypsin
in Earle's balanced salt solution, pH 7.8-MEM (with 10% fetal-
calf serum, and nonessential amino acids) (1:1), centrifuged at
1000 x $\underline{g}$ for 5 min, and suspended in MEM-10% fetal-calf serum
at a density of 10^6 cells/ml. One volume of virus inoculum was
added to 10 vol of mouse-embryo cell suspension. The mixture
was held 1 hr at 23°. Cells were diluted 10 times with MEM-10%
fetal-calf serum, cultured in 6-cm petri dishes at a density of
3.5 x 10^5 cells/dish, and incubated at 37° in a humidified
incubator with 5% CO_2-95% air.

Monolayers were scraped with a rubber policeman, and cells
were suspended in 2-3 ml of supernatant fluid, frozen and thawed
three times, shaken and pipetted vigorously, and homogenized in
a Ten-Broeck grinder, or were sonicated in a Branson sonifier
for 2 min at 6A.

Supernatant fluids for virus pools were decanted 14-16 days after infection, without disruption of the monolayer. Fluid and cell suspensions were stored at -60°.

    d. Murine and Feline Viruses Grown in Human Lymphoid Cultures. The Kirsten strain of mouse sarcoma virus (KiMSV) or the feline leukemia virus (FeLV) was maintained in MEM containing 10% fetal-calf serum and antibiotics (see above) in human adult fibroblast cells. Media containing the virus were clarified at 10,000 x g for 20 min, then the virus was pelleted at 53,000 x g for 2 hr and resuspended to a final concentration 100 times the original. Human lymphoblastoid cells (10^7 viable cells) were suspended in 1.0 ml of the virus suspension and infected on a magnetic stirrer at 37° under CO_2 for 2 hr. Cells were washed twice and suspended in 20 ml of medium to a density of 5 x 10^5 cells/ml. Control cells were passaged weekly.

    e. High Titers of Group-Specific Antigen of Rauscher Leukemia Virus in Rat or Mouse Embryo Cells. Murine leukemia and sarcoma virus strains were serially passaged in NIH mouse-embryo cells, prepared as by Hartley et al. [41]. Fisher rat-embryo cells were prepared in the same medium and cultivated in MEM with 2 mM of glutamine, 100 units of penicillin,and 50 µg of streptomycin per ml, and 10% unheated fetal-calf serum.

    Bottles (500-ml) with mouse- or rat-embryo cells were infected with 10^4 infectious units of Rauscher leukemia virus. Cultures were harvested 14 days after infection. Antigen was

prepared from the cells as described in Rhim et al. [79]. The
microcomplement fixation test was used to assay for the antigen
[43] with antiserum from Fisher rats bearing transplanted sar-
comas induced by Moloney sarcoma virus. Control antigens were
prepared from uninfected rat-embryo cells. Complement fixation
titers were 1:64-1:128. With antigen from uninfected rat-embryo
cells (controls), a titer of 1:8 was obtained. Higher titers
were obtained with Rauscher virus antigens from rat-embryo cells
than from mouse-embryo cells. The same titer was obtained with
murine leukemia (MLV) and sarcoma viruses.

In the microcomplement fixation test, special plates with
wells are used. Amounts of 25 or 50 μl of reagents can be added,
or multiples of these quantities, by special micropipettes.
Dilution is done by special microdilutors which transfer 25 or
50 μl. It is advisable to check the microdilutors frequently
by use of special test paper. When cell suspensions are added,
quick addition of cells is recommended to avoid sedimentation.

f. Preparation from Avian Myeloblastosis Virus. Avian
myeloblastosis virus (AMV) was obtained from the two following
sources by the methods indicated.

(i) Supernatant fluids from infected tissue culture. An
example is chick-embryo fibroblasts of type C/O and C/A [100B].
The C/O cells are susceptible to the A, B, C, and D subgroups
of avian leukosis virus. The C/A cells are resistant to viruses

of subgroup A. Secondary cultures of chick-embryo fibroblasts
were grown in MEM with calf serum.

(ii) Plasma of leukemic chickens. The virus was titrated
by assay of ATPase [66] and stored at -20° to -70°. Tissue-
culture fluids or plasma infected with AMV were centrifuged for
30 min at 3500 x g. The supernatant with virus was concentrated by ultracentrifugation at a performance index of 10 [34].
Pellets were suspended in PBS, pH 7.0. The virus suspension
was layered on a discontinuous gradient of 1.5 ml each of 50,
35, and 20% sucrose in 0.1 M Tris·HCl (pH 7.4) and centrifuged
at 35,000 rpm in an SW39 Spinco rotor for 30 min at 20°. Virus
banded at the middle of the 35% layer. The virus band was
diluted, centrifuged at a performance index of 10, suspended,
layered over a similar sucrose gradient as above, and centrifuged. Virus was then dialyzed at 4° for 12 hr against 40 mM
Tris·HCl, 20 mM sodium acetate, and 1 mM Na_2EDTA (pH 7.2).

The AMV was solubilized by treatment with Tween 80 and
ether, and by extraction twice more of the fatty interphase
with 0.01 M K-phosphate buffer (pH 7.4) with Tween 80
(0.5 mg/ml). Aqueous phases were centrifuged for 30 min at
70,000 x g. The clarified supernatant was concentrated by
ultrafiltration with a Diaflo UM-10 membrane. Sucrose was
added to the concentrated solution to a concentration of 10%.
The antigen-sucrose mixture was layered on a Sephadex G-100

column (2.6 x 26 cm) and eluted with 0.1 M K-phosphate buffer
(pH 7.4) at a rate of 20 ml/hr. Absorbance was read at 260 and
280 nm.

Fractions that possessed gs antigens were pooled, dialyzed
against 15 mM sodium citrate buffer (pH 4.3)-0.02% thiodiglycol,
concentrated by ultrafiltration, and chromatographed on a column
of carboxymethyl (CM)-cellulose (0.9 x 9 cm) equilibrated with
sodium citrate buffer. Proteins were eluted with a linear
gradient of 200 ml of sodium citrate buffer (pH 6.8) in 0.02 M
Na_2HPO_4.

Virus Proteins for Electrophoresis. Purified virus was
suspended in solution with 2.5 mM Mg^{++} and 25 mM K^+, and treated
with 0.5% sodium dodecyl sulfate at 20°. An equal volume of
cold phenol (pH 7.0) with 0.1% hydroxyquinoline and saturated
with 40 mM Tris·HCl—20 mM sodium acetate—1 mM Na_2 EDTA (pH 7.2)
was added. The mixture was shaken for 15 min at 4°. The aqueous
phase and interphase were extracted again with phenol. The
phenol layer was extracted with the Tris—acetate—EDTA buffer.
Phenol and interphase were precipitated with 5 volumes of 99%
ethanol—25% of the volume 20% K acetate, pH 5.

Proteins were recovered from the alcohol phase after 12 hr
at -20° by centrifugation and dissolved in 10% sodium dodecyl
sulfate—0.1 M dithiothreitol. The mixture was incubated at
37° for 10 min, and precipitated with alcohol as described
above. This was repeated, proteins were dissolved in 2 ml of

distilled water, and dialyzed for 3 hr at 20° against twice

concentrated Tris—acetate—EDTA buffer with 0.2% sodium dodecyl

sulfate.

Preparative Gel Electrophoresis.

(i) Gel mixture: 27.22 ml of water, 12 ml of 10 x Tris—

acetate—EDTA, 0.4% N,N'-methylene bisacrylamide, 0.12 ml of

tetramethylethylene diamine.   The mixture was degassed at 15

Torr for 15 min.   1.2 ml of 10% sodium dodecyl sulfate—1 M

dithiothreitol—0.60 ml of 10% ammonium persulfate were added.

The mixture was quickly poured into the gel chamber.

(ii) 5 ml of buffer was immediately overlayed over gel

surface.

(iii) Polymerization was at 20° for 1 hr.

(iv) The gel column was sliced at the lower end and the

elution cell was attached.

(v) 150 mA (100V) was applied for 1 hr, for removal of

interferring material.

(vi) 10% ethylene glycol (5°) was circulated for about

18 hr.

(vii) Sample: 4—22 mg of protein in 2% sodium dodecyl

sulfate-2 x Tris—acetate—EDTA, brought to 7% glycerol.   2.0 ml

of the sample was applied to the column.

(viii) Electrophoresis: 150 mA (100V) for 30 min at 5-8°.

(ix) Elution rate: 20 ml/hr, read at 280 nm.

Analytical Acrylamide Gel Electrophoresis.

(i) Gel solutions: 40 mM Tris·HCl—20 mM sodium acetate —1 mM Na$_2$EDTA (pH 7.2); 10% acrylamide; 0.267% bisacrylamide; 0.1% sodium dodecyl sulfate; 5 mM dithiothreitol.

(ii) Polymerization: 0.05% ammonium persulfate in 10 x 0.6 cm columns.

(iii) Interfering ions: Removed by application of 5 mA/gel for 1 hr in Tris—acetate—EDTA buffer with 0.1% sodium dodecyl sulfate.

(iv) Proteins: Treated with 20 mM dithiothreitol, dissolved in buffer [see (i)] with 7% glycerol.

(v) Samples: 50-400 μg of protein.

(vi) Marker: phenol red.

(vii) Electrophoresis: At 20°, in Tris—acetate—EDTA with 0.1% sodium dodecyl sulfate. 5 mA (75-80 V)/gel.

Gel Staining.

(i) Stain: 0.5% amido black in 7% acetic acid for 2 hr or 0.25% Coomassie Blue in 10% acetic acid—50% methanol.

(ii) Destain with several changes of 7% acetic acid.

(iii) Stain with periodic acid Schiff [110].

(iv) Fix in 12.5% Cl$_3$CCOOH for 30 min, incubate in 3% acetic acid—1% periodic acid for 1 hr. Wash for 19 hr, stain with 0.5% basic fuchsin in the dark for 1 hr. Wash 3 times (10 min each time) with fresh 0.5% potassium metabisulfite, rinse, and store in 7% acetic acid.

g. Antiserum to Group-Specific Antigens. Sera were
obtained from Fisher rats bearing transplanted tumors induced
by Moloney sarcoma virus that possessed large amounts of murine
gs antigen. Only sera with titers of 1:160 or higher were used.
Serum titration was done with antigens from mouse or rat tumor
tissues (10% suspension). Nonspecific antigens were screened
with antigens from uninfected tissues or cultures. Antigens
were tested against 1:2 and 1:4 dilutions of 8 units of rat
serum.

Antigens were tested also by complement fixation and gel
diffusion with 4 units of hyperimmune antiserum, which was prepared in guinea pigs by injection into foot pads twice at 10-day
intervals with purified gs antigen and Freund's adjuvant (2:1).
Four extra injections of antigen without adjuvant were also made
at 10-day intervals. Antisera were highly specific for gs antigen; they did not react with normal cell antigen or with intact
virions in complement-fixation or gel-diffusion tests.

V. CELLS TRANSFORMED BY DNA ONCOGENIC VIRUSES

A. Complement-Fixing and Transplantation Antigens of DNA Oncogenic Viruses

Complement-fixing antigens occur mainly in the virus, and
are located in the nucleus. The immune response to these antigens is humoral (see Tables 3 and 4).

Transplantation antigens occur in the virus and in the tumor, and are located on the cell surface. The immune response to these antigens is cellular (see Table 5).

TABLE 5

Methods for Quantitation of the Cellular Immune Response to
Transplantation Antigens[a] that are Tumor-Specific

Method	References
Neutralization	42, 52, 53, 60, 73, 83, 106, 107
In vitro transformation of lymphocytes	4-6, 28, 46, 69
Target cell, sensitized lymphocytes, and macrophages; interaction in vitro	38, 54, 81, 83
Delayed hypersensitivity	11, 17-19, 22, 23, 33, 56, 61A
Adoptive transfer of immunity	9, 10, 52, 60, 61A, 63-65, 93
In vitro cell migration	33

[a]Tumor-specific transplantation antigens were detected in almost all animal neoplasms tested.

Frequently-used tests to detect the cellular response are (1) detection of delayed hypersensitivity in vitro, (2) isotopic antiglobulin, (3) cytophilic antibody, and (4) antiserum and complement lysis.

1.   Detection of Delayed Hypersensitivity in Vitro

The amount 30 ml of Bayol (Fisher) was injected intraperi-

toneally into guinea pigs, nine days after the injection of
antigen in complete Freund's adjuvant. Peritoneal cells were
harvested 72 hr after the Bayol injection, and cells were washed
three times with minimal essential medium (MEM) and centrifuged
in capillary tubes. Parts of tubes with packed cells were cut
and placed in 0.8-ml Mackaness-type chambers (2 tubes/chamber
[33]). The chambers were filled with dilutions of test antigens or with medium without antigen and incubated at 37° for
24 hr. Cells migrated out of the tubes into the glass chambers.
The area of cell migration was measured.

2. Isotopic Antiglobulin Technique

Target cells were obtained from mice 24 days after inoculation of oncogenic virus (5 x 10^3 tumor-producing units/mouse).
Spleens from infected and normal (control) mice were minced.
Erythrocytes were lysed by addition of 20 mM Tris·HCl-50mM
NH_4Cl buffer, pH 7.8. The spleen cells were labeled with ^{51}Cr
by incubation at 37° for 30 min with 20 µCi of $^{51}Cr/10^8$ cells in
MEM with 15% fetal-calf serum; labeled cells were washed three
times with PBS, pH 7.4. Sera were diluted 1:10 in PBS to give
an excess of antigen.

The quantity 4 x 10^6 cells was added to test tubes with
0.1 ml of serum, and the tubes were shaken gently and incubated
at 24° for 30 min. Cells were washed ten times with PBS at 4°.

Then 10 μl of ^{125}I-labeled antiserum to mouse globulin, prepared in sheep, was added to each tube.

Cells were counted; the ^{51}Cr and ^{125}I were measured simultaneously.

3. Cytophilic Antibody

a. In Vivo Assay.

(i) An amount 0.2 ml of cytophilic antibody was injected with 2-3 x 10^6 tumor cells, intraperitoneally, into mice, 6 hr before they were killed.

(ii) The peritoneal cavity of the mice was washed with MEM, which was centrifuged for 7 min at 1200 x g. The pellet was suspended in MEM supplemented with 0.1 vol of normal mouse serum.

(iii) Wet droplets were prepared on a slide, covered, and examined by phase-contrast microscopy, or by light microscopy after staining with Giemsa. However, phase-contrast microscopy of unstained specimen better reveals the attachment of tumor cells.

(iv) The number of tumor cells attached per macrophage and the percent of macrophages with attached tumor cells were determined. It is also important to observe these morphological changes and to quantitate them in tumor cells and macrophages: vacuolation, size, presence of refractile spherules, clumping,

and/or rosette formation. This is also a test for the integrity
of receptors for cytophilic and opsonic antibodies, as well as
for the presence of cytophilic antibody in a test serum.

    b. In Vitro Assay. The yield of peritoneal cells was
enhanced by intraperitoneal injection of 5% each of starch and
proteose-peptone broth in isotonic saline, four days before the
mice were killed (see in vivo assay). The quantity 1.5×10^6
peritoneal cells (monocytes) was left to adhere for 90 min at
37° to coverslips; monolayers were washed with MEM. Test serum
(0.5 ml) was added, and coverslips with cells and test or normal sera were incubated at 23° for 90 min. Coverslips were
washed again with MEM. Then 3×10^6 tumor cells were added to
each coverslip, which was then incubated at 23° for 90 min.
Coverslips were washed gently by dipping in MEM. Attachment
was examined by phase-contrast microscopy (see in vivo assay).

4. Antiserum and Complement Lysis

    Rabbit antiserum against whole cells was prepared. Cells
in exponential growth were sensitized with heat-inactivated
serum at 37° for 60 min. Cells were washed with PBS, pH 7.6,
and sensitized cells plus complement (0.05%) were incubated at
37° for 30 min. Cytolytic reaction was determined by substitu-
tion of medium with complement, with fresh-growth medium. In
certain experiments, sensitized cells were irradiated before

the addition of complement; controls were similarly treated, except that the cells were unsensitized. Other controls included cells plus antibody alone, and cells plus complement alone.

B. Group-Specific Antigens

These antigens can be detected in virus-free hamster and rat tumors. Tumors induced by adenoviruses type-12 and type-19 do not possess demonstrable infectious adenovirus; however, animals bearing tumors induced by these adenoviruses develop high titers of complement-fixing antibodies that can react with homotypic antigens of tissue culture cells infected with adenovirus. Complement-fixing antibodies can be detected also in hamsters bearing transplants of the noninfectious tumors. Hamsters injected with adenovirus or tumor transplants, but which do not develop tumor growth, possess no complement-fixing antibodies.

1. Inoculum

Adenoviruses type-12 and type-18 grown in KB cells and frozen at -60° were used to infect rats and hamsters, and as antigens for the complement-fixation test. Titers in primary human-embryo kidney cultures kept over a period of 23-30 days were, in $(TCID)_{50}$ (median tissue-culture infective dose) per 0.1 ml, $10^{6.8}$-$10^{7.4}$ for adenovirus type-12 and $10^{5.4}$-10^6 for

type 18. Complement-fixing titers were 1:16-1:32 for type-12
and 1:8-1:16 for type 18.

2. Animals

Minced tumors were inoculated subcutaneously with an 18-
gauge trocar into Syrian hamsters and Sprague-Dawley rats maintained on sterile bedding and cages in quarters free of polyomavirus and SV40. Inoculation subcutaneously is best in case of
tumor induction, since it will be easy to follow tumor growth.

3. Serum

Animals injected with virus were bled 15 days after inoculation; those injected with tumor fragments were bled 30 days
after inoculation. Antigens were evaluated by neutralization
or by immunodiffusion assays with standard antisera (Tables 3-5).

REFERENCES

1. M. Ahmed and G. Schidlovsky, Cancer Res., 32, 187-192
(1972).
2. D. W. Allen, Biochem. Biophys. Acta., 154, 338 (1968).
3. D. W. Allen, D. S. Sarma, H. D. Niall, and R. Sauer,
Proc. Natl. Acad. Sci. U. S., 67, 837-842 (1970).
4. F. Bach and K. Hirschhorn, Science, 143, 813-814 (1964).
5. F. Bach, K. Hirschhorn, R. R. Schreibman, and C. Ripps,
Ann. N. Y. Acad. Sci., 120, 299-302 (1964).
6. F. H. Bach, in Biological Recognition Processes (R. A.
Good and R. T. Smith, eds.), 4th Developmental Immunology Workshop, 1968.

7. H. Bauer and T. Graf, Virology, 37, 157-161 (1969).
8. H. Bauer and D. P. Bolognesi, Virology, 42, 1113-1126 (1970).
9. R. E. Billingham, L. Brent, and P. B. Medawar, Proc. Roy. Soc. (London) B143, 58-80 (1954).
10. R. E. Billingham, W. K. Silvers, and D. B. Wilson, J. Exptl. Med., 118, 397-420 (1963).
11. B. B. Bloom and M. W. Chase, Progr. Allergy, 10, 151-255 (1967).
12. D. P. Bolognersi and H. Bauer, Virology, 42, 1097-1112 (1970).
13. C. W. Boone and K. Blackman, Cancer Res., 32, 1018-1022 (1972).
14. H. B. Bosmann, FEBS Letters, 13, 121-124 (1970).
15. S. V. Boyden, Immunology, 7, 474-483 (1964).
16. S. Breese and T. P. Zacharia, (1973) in Methods in Molecular Biology (J. A. Last and A. Laskin, eds.), Marcel Dekker, New York, in press.
17. L. Brent, J. B. Brown, and P. B. Medawar, Proc. Roy. Soc. (London) B156, 187-209 (1962).
18. M. W. Chase, in The Nature and Significance of the Antibody Response (A. M. Pappenheimer, Jr., ed.), Columbia Univ. Press, New York, 1953, pp. 156-159.
19. W. H. Churchill, H. J. Rap, B. S. Kronman, and T. Borson, J. Natl. Cancer Inst., 41, 13-29 (1968).
20. C. M. Croce, H. Koprowski, and H. Eagle, Proc. Natl. Acad. Sci. U.S., 69, 1953-1956 (1972).
21. J. E. Coe and K. K. Takemoto, J. Natl. Cancer Inst., 49, 39-44 (1972).
22. J. R. David and P. Y. Paterson, J. Exptl. Med., 122, 1161-1171 (1965).
23. J. R. David, Federation Proc., 27, 6-12 (1968).
24. P. J. Deckers and Y. H. Pilch, Cancer Res., 32, 839-846 (1972).
25. R. M. Dougherty and H. S. DiStefano, Virology, 27, 351-354 (1965).
26. W. P. Drake, P. C. Ungaro, and M. R. Mardiney, Cancer Res., 32, 1042-1044 (1972).
27. P. H. Duesberg, H. L. Robinson, W. S. Robinson, P. J. Huebner, and H. C. Turner, Virology, 36, 73-86 (1968).
28. R. W. Dutton, J. Exptl. Med., 122, 759-770 (1965).
29. M. M. Elkind and H. Sutton, Radiation Res., 13, 556-593 (1960).
30. P. J. Fischinger, J. Lange, and W. Schafer, Proc. Natl. Acad. Sci. U.S., 69, 1900-1904 (1972).
31. G. Geering, W. D. Hardy, L. J. Old, E. de Haven, and R. S. Brodey, Virology, 36, 678-707 (1968).
32. G. Geering, A. Tado, and L. J. Old, Nature, 226, 265-266 (1970). G. Geering, T. Aoki, and L. J. Old, Nature, 226, 265-266.

33.   M. George and J. H. Vaughan, Proc. Soc. Exptl. Biol.
Med., 111, 514-521 (1962).
34.   P. Giebler, Z. Naturforsch., 13b, 238-241 (1958).
35.   R. V. Gilden, S. Oroszlan, and R. J. Huebner, Nature
New Biol., 231, 107-108 (1971).
36.   R. V. Gilden and S. Oroszlan, Proc. Natl. Acad. Sci.
U.S., 69, 1021-1025 (1972).
37.   D. M. Goldenberg and H. J. Hansen, Science, 175, 1117-
1118 (1972).
38.   A. Govaerts, J. Immunol., 85, 516-522 (1961).
39.   M. Green, Proc. Natl. Acad. Sci. U.S., 69, 1036-1041
(1972).
40.   K. Hannestad, M.-S. Kao, and H. Eisen, Proc. Natl.
Acad. Sci. U.S., 69, 2295-2299 (1972).
41.   J. W. Hartley, W. P. Rowe, W. I. Capps, and R. J.
Huebner, Proc. Natl. Acad. Sci. U.S., 53, 931-938 (1965).
42.   K. E. Hellstrom and I. Hellstrom, Adv. Cancer Res.,
12, 167-223 (1969).
43.   R. J. Huebner, W. P. Rowe, H. C. Turner, and W. T.
Lane, Proc. Natl. Acad. Sci. U.S., 50, 379-389 (1963).
44.   R. J. Huebner, J. W. Hartley, W. P. Rowe, W. T. Lane,
and W. I. Capps, Proc. Natl. Acad. Sci. U.S., 56, 1164-1169
(1966).
45.   J. Hilgers, R. C. Nowinski, G. Geery, and W. Hardy,
Cancer Res., 32, 98-106 (1972).
46.   K. Hirschhorn, F. Bach, R. L. Kolodny, I. L. Firschein,
and N. Hasham, Science, 142, 1185-1187 (1963).
47.   M. H. Julius, T. Masuda, and L. A. Herzenberg, Proc.
Natl. Acad. Sci. U.S., 69, 1934-1938 (1972).
48.   J. Kieler, C. Radjikowski, J. Moore, and K. Ulrich,
J. Natl. Cancer Inst., 48, 393-405 (1972).
49.   J. B. Kraehenbuhl and J. D. Jamieson, Proc. Natl.
Acad. Sci. U.S., 69, 1771-1775 (1972).
50.   N. K. Jerne, A. A. Nordin, and C. Henry, in Cell-Bound
Antibodies (B. Amos and H. Koprowski, eds.), The Wistar Institute Press, Philadelphia, Pa., 1963, pp. 109-125.
51.   K. Kikuchi, Y. Kikuchi, M. E. Phillips, and C. M.
Southam, Cancer Res., 32, 516-521 (1972).
52.   G. Klein, H. O. Sjogren, E. Klein, and K. E. Hellstrom,
Cancer Res., 20, 1561-1572 (1960).
53.   P. Koldovsky, Folia Biol. (Prague), 7, 115-121 (1961).
54.   H. Koprowski and M. V. Fernandes, J. Exptl. Med., 116,
467-476 (1962).
55.   J. A. Last and T. P. Zacharia, in Methods in Molecular
Biology, Marcel Dekker, New York, in press (1973).
56.   H. S. Lawrence, Ann. Rev. Med., 11, 207-230 (1960).
57.   K. M. Lee, S. Nomura, R. H. Bassin, and P. J. Fischinger, J. Natl. Cancer Inst., 49, 55-60 (1972).

58. L. Levine, in Handbook of Experimental Immunochemistry (D. M. Weir, ed.), Blackwell Scientific Publications, Oxford, England, 1967, pp. 707-719.

59. N. L. Levy, M. J. Mahaley, and E. D. Day, Cancer Res., 32, 477-482 (1972).

60. J. L. McCoy, A. Fefer, N. T. McCoy, and W. H. Kirsten, Cancer Res., 32, 343-349 (1972).

61(A). P. B. Medawar, Harvey Lectures, Ser. 52, 144-176 (1958).

61(B). H. Meier, B. A. Taylor, M. Cherry, and R. J. Huebner, Proc. Natl. Acad. Sci. U.S., 70, 1450-1455 (1973).

62. P. Meyers and R. M. Dougherty, Immunology, 23, 1-6 (1972).

63. N. A. Mitchison, Proc. Roy. Soc. (London), B142, 72-87 (1954).

64. N. A. Mitchison, J. Exptl. Med., 102, 157-177 (1955).

65. N. A. Mitchison and O. L. Dube, J. Exptl. Med., 102, 179-197 (1955).

66. E. B. Mommaerts, D. G. Sharp, J. C. Painter, and J. W. Beard, J. Natl. Cancer Inst., 14, 1011-1025 (1954).

67. P. S. Moorhead, P. C. Nowell, W. J. Mellman, D. M. Battips, and D. A. Hungerford, Exptl. Cell Res., 20, 613 (1960).

68. H. D. Niall, R. Sauer, and D. W. Allen, Proc. Natl. Acad. Sci. U.S., 67, 1804-1809 (1970).

69. G. Pearmain, R. R. Lycette, and P. H. Fitzgerald, Lancet, I, 637-738 (1963).

70. J. C. Petricciani, R. L. Kirschstein, R. E. Wallace, and D. P. Martin, J. Natl. Cancer Inst., 48, 705-713 (1972).

71. M. Pollard, Perspectives in Virology, Wiley, New York, 1959.

72. V. R. Potter, E. F. Baril, M. Watanabe, and E. D. Whittle, Adv. Enzyme Regulation, 8, 299-310 (1970).

73. L. J. Old, E. A. Boyse, D. A. Clarke, and E. A. Carswell, Ann. N. Y. Acad. Sci., 101, 80-106 (1962). L. J. Old, E. A. Boyse, G. Geering, and H. F. Oettgen, Cancer Res., 28, 1288 (1968).

74. S. Oroszlan, C. L. Fisher, T. B. Stanley, and R. V. Gilden, J. Gen. Virol., 8, 1-10 (1970).

75. S. Oroszlan, C. Foreman, G. Kelloff, and R. V. Gilden, Virology, 43, 665-674 (1971).

76. S. Oroszlan, R. J. Huebner, and R. V. Gilden, Proc. Natl. Acad. Sci. U.S., 68, 901-904 (1971).

77. S. Oroszlan, D. Bova, M. H. Martin-White, R. Toni, C. Foreman, and R. V. Gilden, Proc. Natl. Acad. Sci. U.S., 69, 1211-1215 (1972).

78. R. A. Reisfeld, V. J. Lewis, and D. E. Williams, Nature, 195, 281 (1962).

79. J. S. Rhim, L. B. Williams, R. J. Huebner, and H. C. Turner, Cancer Res., 29, 154 (1969).

80. J. S. Rhim and R. J. Huebner, Nature, 226, 646-647
(1970).
81. N. R. Rose, J. H. Kite, T. K. Doebbler, and R. G.
Brown, in Cell-Bound Antibodies (B. Amos and H. Koprowski, eds.),
Wistax Inst. Press, Philadelphia, Pa., 1963, pp. 19-34.
82. S. M. Rose and F. C. Rose, Cancer Res., 12, 1-7 (1952).
83. W. Rosenau and H. D. Moon, J. Natl. Cancer Inst., 27,
471-478 (1961).
84. H. Rubin, Virology, 10, 29-49 (1960).
85. V. Rutishauser, C. F. Millette, and G. M. Edelman,
Proc. Natl. Acad. Sci. U.S., 69, 1596-1600 (1972).
86. P. S. Sarma, H. D. Turner, and R. J. Huebner, Virology,
23, 313-318 (1964).
87. W. P. Park and E. M. Scolnick, Proc. Natl. Acad. Sci.
U.S., 69, 1766-1770 (1972).
88. E. M. Scolnick, W. P. Park, and D. M. Livingston,
J. Immunol., 109, 570-577 (1972).
89. J. T. Sever, J. Immunol., 88, 320 (1962).
90. A. L. Shapiro, E. Vinuela, and J. V. Maizel, Biochem.
Biophys. Res. Commun., 28, 815 (1967).
91. W. V. Shipley, J. Natl. Cancer Inst., 48, 651-655
(1972).
92. R. E. Shope, in Viruses (M. Delbruck, ed.), California
Institute of Technology, Pasadena, 1950, pp. 79-92.
93. G. D. Snell, Ann. Rev. Microbiol., 11, 439-458 (1957).
94. F. T. Steck and H. Rubin, Virology, 29, 628-641 (1966).
95. C. M. Steel, J. Natl. Cancer Inst., 48, 623-628 (1972).
96. W. A. Stenback, G. L. Van Hoosier, and J. J. Trentin,
Proc. Soc. Exptl. Biol. Med., 122, 1219-1221 (1966).
97A. W. A. Stenback, G. L. Van Hoosier, and J. J. Trenton,
J. Virol., 2, 1115-1118 (1968).
97B. S. E. Stewart and B. E. Eddy in Perspectives in
Virology (M. Pollard, ed.), Wiley, New York, pp. 245-255.
98. B. A. Taylor, H. Meier, and D. D. Myers, Proc. Natl.
Acad. Sci. U.S., 68, 3190-3194 (1971).
99. C. C. Ting, D. H. Lavrin, K. K. Takemoto, R. C. Ting,
and R. B. Herberman, Cancer Res., 32, 1-6 (1972).
100A. C. C. Ting, D. H. Lavrin, G. Shiu, and R. B. Herberman, Proc. Natl. Acad. Sci. U.S., 69, 1664-1668 (1972).
100B. P. K. Vogt and R. Ishizaki, Virology, 26, 664-672
(1965).
101. G. Walter, R. Roblin, and R. Dulbecco, Proc. Natl.
Acad. Sci. U.S., 69, 921-924 (1972).
102. K. Weber and M. Osborn, J. Biol. Chem., 244, 4406-
4409 (1969).
103. H. Welcerle, P. Lonai, and M. Feldman, Proc. Natl.
Acad. Sci. U.S., 69, 1620-1624 (1972).
104. R. Werner and D. Moon, J. Natl. Cancer Inst., 27,
471-483 (1961).

105. E. F. Wheelock, S. T. Toy, W. L. Caroline, L. R.
Sibel, M. A. Fink, P. C. L. Beverley, and A. C. Allison, J.
Natl. Cancer Inst., 48, 665-673 (1972).

106. H. J. Winn, Ann. N. Y. Acad. Sci., 101, 23-45 (1962).

107. H. J. Winn, Natl. Cancer Inst. Monograph, 2, 113-138
(1959).

108. T. P. Zacharia and M. Pollard, J. Natl. Cancer Inst.,
35-41 (1969).

109. T. P. Zacharia, J. Reticuloendothel. Soc., 9, 138-146
(1971).

110. R. M. Zacharius and E. Zell, Anal. Biochem., 30, 148-
152 (1969).

Chapter 6

METHODS FOR THE STUDY OF THE CELLULAR BASIS OF IMMUNOLOGICAL TOLERANCE

Gail S. Habicht

Department of Microbiology, School of Basic Health Sciences
State University of New York at Stony Brook
Stony Brook, New York

Jacques M. Chiller and William O. Weigle

Department of Experimental Pathology
Scripps Clinic and Research Foundation
La Jolla, California

I.   INTRODUCTION 142

II. METHODS APPLICABLE TO THE STUDY OF THE CELLULAR
     BASIS OF IMMUNOLOGICAL TOLERANCE 143

     A. Lymphoid-Cell Transfer 145
     B. Plaque-Forming-Cell (PFC) Assay 152
     C. Conjugation of Serum Proteins to Erythrocytes . . 154
     D. Preparation of a Tolerogen and the
         Corresponding Immunogen 155
     E. Detection of Specific Antigen Receptors
         on Lymphoid Cells 157

     REFERENCES 159

I. INTRODUCTION

When lymphoid tissues of an animal are exposed to an antigen they may respond in different ways. The host may produce
humoral antibody of one or more immunoglobulin class, it may
develop delayed hypersensitivity, or it may become specifically
tolerant to future immunogenic challenge with that antigen. An
animal is said to be tolerant when, as a result of previous
exposure to an antigen, it is no longer able to produce a detectable immunity to that antigen. It is likely that the mechanisms for induced immunological tolerance are similar to the
physiological mechanisms for maintaining tolerance to self constituents [1].

The cellular basis for immunological tolerance has been the
subject of much speculation and experimentation. Central questions that remain to be answered are whether or not there are
specific immunologically unresponsive cells and what types of
lymphoid cells are involved. Is an animal rendered tolerant by
the elimination of specific immunocompetent clones of cells, or
are specific clones of cells merely suppressed? If cells are
suppressed rather than killed, are they suppressed by antigen
or by a product elaborated by another cell type?

II. METHODS APPLICABLE TO THE STUDY OF THE CELLULAR BASIS
OF IMMUNOLOGICAL TOLERANCE

Certain experimental techniques have been used repeatedly in elucidation of the cellular basis of the immune response, and hence, are applicable to the study of the cellular basis of immunological tolerance. The technique developed by Jerne and Nordin [2] for visualizing single antibody-forming cells to sheep erythrocytes has been modified for use with protein [3], polysaccharide [4], and haptenic [5] antigens. The immune response quantitated by this technique is the number of individual antibody-forming cells, known as the plaque-forming-cell (PFC) response. Since it is difficult to study the cellular basis of immunological tolerance in the animal per se, many experiments have involved reconstitution of the immune response in an immunologically neutral environment. These techniques involve harvesting single-cell suspensions from individual lymphoid tissues such as spleen (S), thymus (T), bone marrow (BM), thoracic duct lymphocytes (TDL), or lymph nodes (LN). Ideally, reconstitution experiments should be performed in vitro but with many antigens it is difficult or, until now, impossible to induce a primary immunological response--either antibody formation or tolerance. However, some antigenic systems can be investigated in vitro, for example, the induction of the immune response to heterologous erythrocytes [6] and the induction of tolerance to flagellin [7].

When the reconstitution experiment cannot be performed in
vitro, it is important to provide a background as immunologically
neutral as possible in vivo. The immune response in lethally
irradiated hosts can be reconstituted with syngeneic lymphoid
cells. Reconstitution experiments may or may not require the
presence of T cells, depending on the antigen involved. Some
antigens are referred to as thymus-dependent antigens because
they require the presence of T cells in addition to B cells for
the production of an immune response [8]. There is a synergis-
tic action between these cell types whereby the production of
specific antibody by B cells is enhanced by a helper function
performed by T cells. Reconstitution of the response to thymus-
dependent antigens requires the transfer of immunocompetent T
and B cells or of lymphoid tissues such as the spleen, which is
a composite of both T and B cells. On the other hand, some
antigens are thymus-independent, that is, they can elicit an
antibody response with B cells alone. In order to obtain a
population of B cells devoid of T cells in which thymus dependency can be tested, the following type of reconstitution procedure is performed. The host animal is thymectomized as an adult.
Later, usually after a few weeks, the host is lethally irradiated and then reconstituted with BM cells from a syngeneic donor.
The BM cells are treated with anti-Θ antiserum and complement
(C) in order to destroy any Θ-bearing cells [9]; Θ is an allo-
antigen found on T cells, but apparently not on any other

lymphoid cells [10]. An animal that has been subjected to this procedure is called an adult thymectomized, irradiated, bone-marrow-reconstituted host and may be challenged with an antigen to test for thymus dependency or to study the immunological activities of B cells in the absence of T cells.

A. Lymphoid-Cell Transfer

Lymphoid-cell donors, either normal or tolerant, are decapitated with a guillotine high in the cervical region, with maximum blood loss and no anesthesia. Blood loss is desirable to reduce the number of erythrocytes and circulating lymphocytes in the tissue harvested.

1. Thymus

We take the thymus by opening the thoracic cavity and carefully dissecting away the white thymus tissue. Careful dissection of the two lobes of the thymus from the surrounding tissue will avoid taking the parathymic lymph nodes, whereas cutting out the thymus usually induces bleeding and increases the risk of taking lymph nodes. Tissues removed from the donor are placed in a petri dish of cold (4°C on ice) balanced salt solution (BSS); in milligrams per liter of distilled water

dextrose, 1000	phenol red, 10	NaCl, 8000
KH_2PO_4, 60	$CaCl_2 \cdot 2H_2O$, 186	$MgCl_2 \cdot 6H_2O$, 200
$Na_2HPO_4 \cdot 7H_2O$, 358	KCl, 400	$MgSO_4 \cdot 7H_2O$, 200.

The BSS used in the preparation of cells for transfer contained
100 μg streptomycin and 100 units of penicillin per milliliter.

2. Spleen

Spleens are removed from peritoneal cavities of mice,
excess fat is dissected away, and spleens are placed in petri
dishes with cold BSS.

Single-cell suspensions of both spleen and thymus are
obtained in the same manner. The tissue is grasped in a forceps and rubbed against a 2-cm square of stainless-steel screen
(350-μm mesh) held over the petri dish. This procedure is continued until a small greyish mass of connective tissue is all
that remains. The screen is washed with cold BSS. The resulting suspension is filtered through a nylon cloth (150-μm mesh)
into a conical tube and centrifuged at 225 x g for 10 min. The
supernatant is decanted and fresh BSS is added; the cells are
suspended and centrifuged again. This wash is repeated one or
more times. Cells to be transferred to an irradiated host must
be refiltered through nylon cloth to assure a single-cell suspension. About 80-100 x 10^6 thymus cells and 100-200 x 10^6
spleen cells may be harvested in this way from a normal adult
mouse.

For transfer of splenic and thymic lymphoid cells, the
desired number of cells (90 x 10^6 thymus cells or 100 x 10^6

spleen cells) is suspended in about 0.5 ml. This amount is injected intravenously via a lateral caudal vein. Care must be taken when injecting thymus and spleen cells since rapid injection often causes death. As long as 4 min may be necessary for the infusion of 90 x 10^6 thymus cells.

3. Bone Marrow

Bone marrow is harvested from the femur and the tibia. Legs are skinned and removed from the animal by sectioning above the femur head. Most of the muscle tissue is removed. The femur is grasped in a forceps and the shaft is cut just below the head and above the knee joint. Cold BSS from a 10-ml syringe fitted with a 25-gauge needle is forced through the shaft and the bone marrow is extruded, often as a single piece, from the opposite end into a petri dish on ice. The procedure is repeated for the tibia. A single-cell suspension is prepared by gently drawing up and extruding the bone marrow and BSS mixture first through a syringe with no needle then once through a 19-gauge needle. The cells are filtered and washed as described above. About 30 x 10^6 lymphoid cells per donor (4 bones) may be obtained with this procedure.

For the transfer of bone-marrow cells, the desired number of cells is suspended in 0.2 ml and is injected intravenously.

4. B Cells

The B cells are bone-marrow-derived anti-Θ serum-resistant
cells that are harvested from spleens of adult thymectomized
lethally irradiated bone-marrow-reconstituted donors.

a. Adult Thymectomy. Adult thymectomy is performed as
described by Miller [11]. Adult mice aged 6 to 8 weeks are
anesthetized with nembutal and placed with their dorsal sides
down and their legs stretched laterally. A pad of cotton wool
is used to arch the back, and the head is pulled forward by a
thread attached to the incisors. The skin over the sternum is
cut and a small piece of the sternum (about 2 x 5 mm) is removed
in the midline. Muscles and fascia ventral to the thymus are
cut. Thymus lobes are forced to the surface by pressure on the
abdomen. Each lobe is removed individually by suction. Pressure is exerted on the abdomen until the back-pad and front-leg
retractors are removed and the skin is apposed. The skin inci-
sion is closed with 2 Michel clips. If the operation is performed rapidly, the mice do not suffer from pneumothorax.

b. Lethal Irradiation. Lethal irradiation (900 R) is
given two to four weeks after the adult thymectomy.

c. Bone-marrow reconstitution. Within 3 hr of irradiation,
5 x 10^6 syngeneic bone-marrow cells are injected intravenously.

The B cells are harvested from the spleens of animals that

show no evidence of thymus remnants at the time of killing.
Thirty days after the reconstitution, spleen cells are collected
as described above, then treated with anti-Θ serum (see below).
About 10^9 spleen cells are incubated for 35 min at 37°C with
5-10 ml of a final concentration of anti-Θ serum in BSS that
is 10 times as great as the greatest dilution of anti-Θ serum
that will kill 95% of Θ-bearing thymus cells. After the incubation, BSS is added to a final volume of 50 ml. The mixture
is centrifuged, the supernatant is aspirated, and guinea-pig
complement (C) is added. The C must be absorbed with agarose
to remove any guinea pig antibodies specific for mouse tissue
[12]. To the packed cells, 10 ml of absorbed C diluted 1:10
with BSS is added. The mixture is incubated for 45 min at
37°C and brought to 50 ml with BSS; the cells are washed with
excess BSS. These cells are now Θ-free splenic B cells.

5. Preparation of Anti-Θ Serum

Anti-Θ serum is prepared by injection of Θ-bearing thymus
cells from C_3H mice into Θ-deficient mice of the AKR strain.
Many different protocols are available for production of anti-Θ
antibodies; this one has been successful in this laboratory.
Adult AKR mice are injected at weekly intervals with freshly
harvested C_3H thymus cells as follows:

Week	Cell numbers	Route
1	40 x 10^6	Subcutaneous
2	60 x 10^6	Subcutaneous
3	100 x 10^6	Subcutaneous
4	100 x 10^6	Intraperitoneal
5	100 x 10^6	Intraperitoneal
6	100 x 10^6	Intraperitoneal

One week after the last injection, the animals are bled and individual sera are tested for their anti-θ titer. High-titer sera are pooled for use. Immunized animals may be bled at weekly intervals for at least three weeks.

a. Titration of Anti-θ Serum. The highest dilution (the titer) of the antiserum that will kill 95% of 5 x 10^6 C_3H thymus cells is measured by the following ^{51}Cr-release assay.

b. Labeling of T Cells with ^{51}Cr. Washed T cells are brought to a final concentration of 3 x 10^8 cells/ml in HEPES (N-2-hydroxyethyl piperazine-N-2'-ethane-sulfonic acid)-buffered Eagle's medium, pH 7.2 (Flow Laboratories, Inglewood, Calif.), mixed with an equal volume of ^{51}Cr (120 μCi/ml in saline), and incubated at 37°C for 35 min. The incubation mixture is transferred to a large tube (50 ml) and the cells are washed twice in BSS. The cells are brought to a final concentration of 40-60 x 10^6 cells/ml. The amount 0.5 ml of this mixture should contain about 80,000 cpm.

c. Absorption of C. Normal guinea pig serum is diluted 1:3 in saline; for every ml of diluted C, 20 mg of agarose is added. This mixture is put on ice for 30 min and shaken every

3 min. The mixture is centrifuged at 2250 x g for 10 min, and

the C is carefully aspirated from the agarose.

    d. ^{51}Cr-Release Assay. Starting with 50 µl of each serum

sample, 1:10, 1:100, and 1:500 dilutions are prepared. The fol-

lowing mixture is placed in small plastic precipitin tubes:

0.1 ml of each dilution of each serum; 0.1 ml of ^{51}Cr-labeled

T cells; and 0.1 ml of medium (HEPES). This mixture is incu-

bated for 35 min at 37°C. The tubes are shaken twice during

the incubation and centrifuged at 225 x g for 5 min. The super-

natant is carefully removed without removing any cells. Then

0.1 ml of the absorbed C and 0.2 ml of HEPES are added to the

cells and mixed gently, but thoroughly, with a Pasteur pipette.

The mixture is incubated for 45 min at 37°C and mixed three times

during the incubation. After centrifugation at 225 x g for 10

min, 200 µl of the supernatant is carefully removed and placed

into a small tube for counting. The following controls must

also be run: duplicate tubes with C but no antiserum, dupli-

cate tubes with normal AKR serum instead of antiserum, and

duplicate tubes with labeled cells that have been subjected to

freezing and thawing three times. The results are expressed as

the percent of the ^{51}Cr label released by incubation with the

antiserum, as compared with the amount released by freeze-

thawing.

B. Plaque-Forming-Cell (PFC) Assay

In order to determine the number of individual antibody-
forming cells in a spleen, a single-cell suspension is prepared
from each spleen, as described above. Just after the first filtration through the nylon cloth, the volume of filtrate is noted
(about 12 ml) and a sample is taken for a lymphoid cell count.
This count allows a calculation of the total number of lymphoid
cells per spleen. After one wash with BSS as described above,
the cells are suspended in 5 ml of 0.83% NH_4Cl for 5 min at room
temperature in order to lyse the erythrocytes. The mixture is
centrifuged, then washed at least twice in cold BSS. After the
final wash, the volume is adjusted so that there are about
15×10^6 cells/ml. A cell count is taken (up to 50% of the
splenocytes may be lost by this time). Then 1:10 dilutions of
this suspension are prepared. The number of dilutions prepared
depends on the number of PFC expected, as will be seen below.

For the PFC assay per se the requirements include the following: a water bath at 45°C, a liquid solution of 0.5% agarose
A37 in BSS, a 6-7% suspension of target erythrocytes, the splenic lymphocytes prepared above, microscope slides (which have
been previously coated with 0.1% agarose in water and air dried),
and the necessary micropipettes.

Hot agarose (0.5 ml) in BSS is pipetted into disposable
culture tubes (12 x 75 mm), which are already in the water bath.

The temperature of the water bath is important because 45° main-
tains the agarose in the fluid state and causes minimum cell
damage during a short exposure. Then 50 µl of the target-cell
suspension is added, followed by 100 µl of the splenocyte dilution. The contents of the tube are mixed and immediately poured
onto the clear half of the coated, frosted microscope slide.
The slide is placed on a rack in a moist chamber (bread box with
wet paper towels); when all the slides for the experiment have
been poured, the box is placed in a 37° incubator for 1 hr.

After 1 hr, the C and, in the indirect PFC assay, amplifying antibody are added. The racks are designed so that they
have a trough into which the complement (1:10 dilution of normal guinea pig serum previously absorbed with the target erythrocytes) may be introduced. The slides are inverted so that
the agarose layer hangs into the complement solution. Since
IgG is a hemolytically inefficient antibody, an amplifying antibody must be added to the complement solution when it is desired
to measure the IgG PFC response. This is a previously determined amount of antibody from another species (goat or monkey)
directed against mouse IgG. Similarly, mouse IgA PFC responses
may be detected with anti-IgA as the amplifying serum. The PFC
that are developed with an amplifying antiserum are referred to
as indirect PFC. The ability to measure indirect PFC has been
important for the study of the cellular basis of tolerance to
serum proteins.

The slides inverted in complement, with or without amplifying antibody, are returned in the moist chamber to the 37°C incubator for 2-3 hr. For counting, the optimum number of PFC per
slide is 50-100. If the expected number of PFC/10^6 lymphoid
cells is 10, then one dilution of splenocytes would suffice,
since about 1.5 x 10^6 cells would be added to each slide. However, if 1000 PFC/10^6 lymphoid cells is estimated, then three
dilutions (1:10) would be necessary. The number of lymphoid
cells on the slide from the third dilution would be 1.5 x 10^4,
and 15 PFC could be expected per slide. In order to express the
PFC per spleen, the number is calculated based on the number of
cells per spleen obtained from the first lymphoid cell count.

C. Conjugation of Serum Proteins to Erythrocytes

The original hemolytic plaque assay has been modified for
use with proteins by Golub et al. [3]. These antigens have been
conjugated to either heterologous or homologous erythrocytes,
which are then used as target cells in the PFC assay. Whenever
heterologous erythrocytes are used, a background PFC response
to them must be measured and subtracted from the response to
the antigen conjugated to the cell. A method for the conjugation of human gamma globulin (HGG) to goat erythrocytes will be
discussed as an example. Erythrocytes are obtained and stored
in sterile Alsever's solution. The cells are washed at least

three times in conjugation buffer (in grams/liter of distilled water: NaCl, 4.45; KH_2PO_4, 2.4; Na_2HPO_4, 10.0, pH 7.2). A solution of 20 mg/ml of HGG (Cohn fraction II) in conjugation buffer is absorbed with the target red cells by addition of 9 volumes of HGG solution to 1 volume of packed cells and incubation at 4°C for 1 hr. Cells are centrifuged and the supernatant is harvested. It may be necessary to repeat this step in order to remove all of the hemolytic and agglutinating antibodies in the HGG that are specific for target cells. For the conjugation, 15 ml of HGG (20 mg/ml) is mixed with 0.5 ml of packed goat erythrocytes, then 2.5 ml of freshly prepared 1-ethyl-3-(3-dimethylaminopropyl)carbodiimide·HCl (100 mg/ml in conjugation buffer) is added. The reaction is performed at 4°C for 1 hr with occasional stirring. The conjugated erythrocytes are washed in conjugation buffer and may be stored in this buffer, preferably as the pellet formed by centrifugation. For use in the PFC assay, the conjugated cells are suspended in BSS.

D. Preparation of a Tolerogen and the Corresponding Immunogen

The HGG is an excellent antigen for the study of the cellular basis of immunological tolerance, since it may be obtained in two antigenically identical but biologically distinct forms --one is tolerogenic and the other is immunogenic [13].

1. Tolerogen

The HGG (Fraction II) is further purified by DEAE-cellulose column chromatography in 0.01 M phosphate buffer (pH 8.0). The purified HGG is further treated to obtain the monomeric deaggregated HGG by ultracentrifugation. A solution of HGG (30 mg/ml) is spun at 40,000 rpm for 150 min in a SW 50.1 swinging bucket rotor in a Spinco L-4 preparative ultracentrifuge (Beckman Industries, Inc., Fullerton, Calif.). The upper third of the material is removed from each tube and diluted to 2.5 mg/ml with 0.15 M NaCl. Then 1 ml is injected intraperitoneally into a mouse within 1 hr of the completion of the ultracentrifugation. Care must be taken to avoid bubbles in the solution, and to avoid shearing forces created by forcing the solution rapidly through a small-gauge needle. These conditions tend to cause a denaturation and reaggregation of the HGG.

2. Immunogen

The immunogenic form of HGG is prepared by heat-aggregation of the HGG purified on DEAE-cellulose, largely as described by Gamble [14]. A solution of HGG (20 mg/ml in 0.01 M sodium phosphate buffer, pH 8.0) is heated at 63°C for 25 min with occasional stirring. After heating, it is placed in an ice bath for 12 hr. The aggregated HGG is precipitated by addition of 2.18 M Na_2SO_4 to a final concentration of 0.62 M. The mixture is put on ice for 20-30 min. The resulting precipitate is washed three

times with 0.62 M Na_2SO_4, then dissolved in phosphate-buffered

saline (PBS), pH 8.0. It is dialyzed free of Na_2SO_4 in PBS and

is diluted to 2 mg/ml for use and stored at -20°C. The usual

immunogenic dose of aggregated HGG is 0.4 mg given intravenously

or intraperitoneally.

E. Detection of Specific Antigen Receptors on Lymphoid Cells

1. Antigen-Binding Cells

Louis et al. [15] have looked for bone-marrow, thymus, and

spleen cells with specific receptors for ^{125}I-labeled HGG.

a. Labeling of HGG. The HGG is iodinated by the chlora-

mine T method [16]. The amount 2-4 mCi of ^{125}I in 50 µl is

neutralized to pH 7.0 with 25 µl of 0.1 M NaH_2PO_4, and 40 µl of

a solution of HGG (250 µg/ml) plus 5 µl of chloramine T (5 mg/ml)

are added; the mixture is allowed to react for 5 min. Then

5 µl of sodium metabisulfite (5 mg/ml), 1 drop of 1% KI, and

0.5 ml of fetal-calf serum are added to the mixture, which is

then passed over a Sephadex G-25 column to remove any free

iodide.

b. Lymphoid Cells. Cells are harvested as described

above and then suspended in minimal essential medium containing

10% fetal-calf serum.

c. Antigen Binding. Lymphoid cells (20 x 10^6) are suspended in 200 µl of minimal essential medium with 10% fetal-calf
serum and 15 mM sodium azide. The sodium azide inhibits the
pinocytosis of antigen by macrophages. The amount 10 µl containing 100 ng of ^{125}I-labeled HGG (50-70 µCi/µg) is added to the
cell suspension, and the mixture is incubated at 4°C for 30 min.
Cells are then layered over 5 ml of fetal-calf serum and are
harvested by centrifugation at 180 x g for 10 min. This last
step is repeated three times.

d. Radioautography. Cells are smeared on methanol-cleaned
microscope slides, fixed in 1% glutaraldehyde, washed in water,
and air dried. Smears are dipped in NTB$_2$ (Kodak) emulsion and
exposed for one month before developing. Cells are stained
with Giemsa after development of the emulsion. Morphologically
intact lymphoid cells with 15 or more silver grains directly
over the cell are considered to be antigen-binding cells.

2. Rosette-Forming Cells (RFC)

Antigen-binding cells have been detected by another assay
in which erythrocytes, either homologous or heterologous, are
coated with antigen. When the sensitized erythrocytes are
mixed with specific receptor-bearing lymphocytes, the erythrocytes may adhere to a lymphocyte and form a rosette consisting
of a cluster of erythrocytes around the lymphocyte. Howard
et al. [17] have used the following method to show that

pneumococcal polysaccharide type III (SIII) RFC are found in

mice tolerant to SIII.   Spleen-cell suspensions and splenic

lymphoid-cell counts are made as described above.   The suspen-

sion is adjusted to a final concentration of 12 x 10^6 cells/ml.

Then 0.5 ml is placed in a glass tube (0.9 cm i.d. x 7.5 cm)

along with 0.5 ml of a suspension of sensitized homologous

erythrocytes (30 x 10^6 cells/ml) and 0.05 ml of normal homolo-

gous serum.   The tube is sealed, centrifuged at 150 x $\underline{g}$ for

7 min, and incubated on ice (4°C) for 90-120 min.   Cells are then

suspended by horizontal rotation for 10 min at 16 rpm on a wheel

of 24 cm radius.   The presence of rosettes is determined micro-

scopically.   At the same time, total nucleated cell counts are

made so that the number of RFC/10^6 nucleated cells may be cal-

culated and, by reference back to the total splenic lymphoid-

cell count, the total RFC/spleen may be calculated.

REFERENCES

1.   F. M. Burnet and T. Fenner, The Production of Antibodies, Macmillan and Co., Ltd., Melbourne, Australia, 1949.
2.   N. K. Jerne and A. A. Nordin, Science, 140, 405 (1963).
3.   E. S. Golub, R. I. Mishell, W. O. Weigle, and R. W.
Dutton, J. Immunol., 100, 133 (1968).
4.   J. G. Howard, G. H. Christie, and B. M. Courtenay,
Proc. Roy. Soc., B178, 417 (1971).
5.   H. Yamada and A. Yamada, J. Immunol., 103, 357 (1969).
6.   R. I. Mishell and R. W. Dutton, J. Exptl. Med., 126,
423 (1967).
7.   E. Diener and W. D. Armstrong, J. Exptl. Med., 129,
591 (1969).

8. H. N. Claman, E. A. Chaperon, and R. F. Triplett, Proc. Soc. Exptl. Biol. Med., 122, 1167 (1966).

9. M. C. Raff, Transplant. Rev., 6, 52 (1971).

10. A. E. Reif and J. M. V. Allen, J. Exptl. Med., 120, 413 (1964).

11. J. F. A. P. Miller, Brit. J. Cancer, 10, 431 (1960).

12. A. Cohen and M. Schlesinger, Transplantation, 10, 130 (1970).

13. J. M. Chiller and W. O. Weigle, J. Immunol., 106, 1647 (1971).

14. C. N. Gamble, Int. Arch. Allergy Appl. Immunol., 30, 446 (1966).

15. J. Louis, J. M. Chiller, and W. O. Weigle, J. Exptl. Med., 137, 461 (1973).

16. P. J. McConahey and F. J. Dixon, Int. Arch. Allergy Appl. Immunol., 29, 185 (1966).

17. J. G. Howard, J. Elson, G. H. Christie, and R. G. Kinsky, Clin. Exptl. Immunol., 4, 41 (1969).

Chapter 7

COMPLEMENT COMPONENTS: ASSAYS, PURIFICATION,
AND ULTRASTRUCTURE METHODS

R. M. Stroud

Medical College of Alabama
Birmingham, Alabama

H. S. Shin

Johns Hopkins School of Medicine
Baltimore, Maryland

and

E. Shelton

National Cancer Institute
National Institutes of Health
Bethesda, Maryland

I.   DESCRIPTION OF THE COMPLEMENT (C) SYSTEM 162

II. HEMOLYTIC ASSAYS 165

     A. Buffers 167
     B. Preparation of Cellular Intermediates 169

C. Choice of Antibody 171
D. CH50 Assay 172
E. C1 Titration 173

III. PURIFICATION OF COMPLEMENT COMPONENTS AND OTHER
      PROTEINS THAT INTERACT WITH THE COMPLEMENT SYSTEM . 177

A. Purification of C3 177
B. Purification of C2 by Specific
    Affinity to EAC4 181
C. Other Complement Components and Some Proteins
    That Interact with Complement 183

IV. BIOLOGICAL ASSAYS 184

A. Bioassay for Anaphylatoxins 184
B. Assay for Leukocyte Chemotaxis 186
C. Immune Adherence 190
D. Phagocytosis 191

V. ULTRASTRUCTURE OF C1q AND LESIONS IN MEMBRANES
    PRODUCED BY THE INTERACTION OF ANTIBODY AND
    COMPLEMENT.   SPECIMEN PREPARATION AND INTER-
    PRETATION OF RESULTS 194

A. Support Films 196
B. Stains 197
C. Optimum Concentration of Substances
    to Be Stained 198
D. Preparation of Membrane Lesions 199
E. Preparation of C1q 201
F. Staining Procedure 202
G. Interpretation of Results 203

ACKNOWLEDGMENT 205

REFERENCES 205

I. DESCRIPTION OF THE COMPLEMENT (C) SYSTEM

Complement is the activity of fresh normal serum that
brings about lysis of cells sensitized with antibody. This

activity is the result of nine sequentially reacting serum proteins that undergo enzyme-substrate reactions generating media-
tors of inflammation and cell damage. The reaction steps that
lead to the assembly of macromolecular enzymes from these nine
proteins take place most efficiently on membrane surfaces. The
final step of cell lysis undoubtedly requires precise spatial
arrangements between the complement components and cell membrane
constituents. Recent reviews of complement functions and the
individual biochemical events should be consulted for more
detailed information [1-3]. The items selected here are chosen
for their current interest.

Complement components are designated C1, C4, C2, C3, C5,
C6, C7, C8, and C9, in the order of their reactivity in lysing
sensitized cells (EA). Component C1 is a complex of three
reversibly interacting protein subcomponents, C1q, C1r, and C1s.
These eleven proteins occur in serum in concentrations ranging
from 20 to 1500 μg/ml. Individual reaction steps in immune
lysis are outlined by the vertical pathway in Fig. 1. The biological functions generated after certain reaction steps are
also shown. According to convention, complement fragments are
designated by the symbol for the component followed by the
small letter a or b, etc., e.g., C3a or C3b. This designation
is generally dropped when the context makes it clear, e.g.,
$\overline{C4b2a}$ could be written $\overline{C42}$. The line over the symbols indicates
that the component or complex is in an active enzyme form, having

been converted from a precursor form. The scheme is usually
outlined as it reacts with a cell-bound antigenic site (S)
sensitized with antibody (SA), although some reaction steps can
take place in the fluid phase without cells and, moreover, without antibody. As diagramed, the interaction of C1q, a subcomponent of C1, with antibody leads to an internal activation of
C1s, which then reacts with its substrates C4 and C2 [4,5].
These two molecules undergo enzymatic cleavage and a new enzyme
$C\overline{42}$ is formed from the larger fragments C4b and C2a. The other
fragments may be involved in kinin formation [6]. The $C\overline{42}$ is a
macromolecular enzyme complex with a half-life of 8 min at 37°C.
This complex brings about the enzymatic cleavage of C3 into C3a
and C3b [7]. The major fragment C3b is incorporated into the
$C\overline{42}$ complex, now designated $C\overline{423}$, and, on reaction with C5,
cleavage of C5 ensues [8]. After C6, C7, C8, and C9 react,
a "lesion," designated S*, is formed and membrane disruption
occurs. Whether or not a cleavage of C6, C7, C8, or C9 occurs
as it does for C4, C2, C3, and C5 is unknown. It is important
to note that there are several naturally occurring inhibitors
of various complement reaction steps (C1s inhibitor, C3 inactivator, C6 inhibitor) and at least one inactivator of two of the
important low-molecular-weight cleavage products C3a and C5a,
the anaphylatoxin inactivator [9]. Other inhibitors may exist
for the homeostatic control of this labile and easily activated
(perhaps continuously activated) system.

Complement components from guinea pigs and humans have received thorough study, and most have been characterized with regard to their approximate molecular weight and electrophoretic mobility in relation to other serum proteins. The available data have been summarized recently for each of these species [10].

A sequential activation of the complement components can be initiated by substances other than immune complexes, namely, aggregates of immunoglobulins [11], complexes of staphlycoccus protein A with immunoglobulins [12], certain polynucleotides [13], denatured insulin aggregates [14], and the saltlike complex of denatured DNA and lysozyme [15]. The late complement components C3 through C9 can be activated by a factor purified from cobra venom, certain gram positive and gram negative organisms, yeast, and various carbohydrate substances [16,17]. For this reaction serum cofactors other than C1, C4, and C2 are needed. One of the cofactors, called factor B in the properdin system (other names are: serum cofactor for cobra venom, glycine-rich beta glycoprotein, and C3 proactivator), has been purified (see Table 1). This factor is heat-labile and requires Mg^{++} for its reaction.

II. HEMOLYTIC ASSAYS

There are four general categories of C-component assays:

(1) hemolytic, (2) immunochemical by the Mancini technique,
(3) neutralization of activity by antibody [18], and (4) relatively nonspecific biological tests both in vivo and in vitro
(e.g., chemotaxis, smooth muscle contraction, and capillary
permeability). Of these, the hemolytic methods have extreme
sensitivity without any loss of specificity. The immunochemical
tests do not distinguish native forms of C components from fragmented forms, nor do they reveal the state of activation unless
carefully prepared antisera are used to study the fragments of
C3 [19]. Immunochemical assays have been useful in clinical
laboratories and in elucidating genetic deficiencies of complement components [20].

An underlying basic concept in immune hemolysis, the one-
hit theory of complement reactivity [21], states that a single
lesion S* formed by the interaction of all nine complement components is sufficient for lysis of a single cell. With proper
control of the assay conditions, it has been possible to lyse
sheep erythrocytes with one to two molecules of C1 [22] or C2
[23] and directly support the theory. However, it is known that
certain components, such as C3, can be bound to cells without
subsequent S* formation (so-called nonproductive sites).

Although the chemical basis of lysis is poorly understood
[24,25], immune hemolysis has been used to develop extraordi-
narily sensitive assays for each of the nine complement components [26-29]. These assays have found clinical and experimental

applications to purified components and complex inflammatory
exudates, and in certain cases have been adapted to serum [30].
In the following discussion, the preparation of buffers, cellular intermediates, and antibody will be applicable to most of
the complement components. Reference will be made to the CH50
assay and, as a prototype assay, the C1 titration will be
described in detail. Other methods are found in the references
cited in Table 1 and many of these are collected in the detailed
book by Rapp and Borsos [31].

A. Buffers

Buffers used for C titrations generally have an ionic
strength that is physiological or partially (0.4 to 0.5) physiological (low-ionic-strength buffer). The osmotic strength is
kept isotonic with sucrose or glucose to avoid cell alterations.
Concentrated stock solutions are used to prepare fresh buffers
each day. In order to check the final ionic strength, the
resistance of the diluted buffer stock at 0°C is determined.
In adjusting various buffers and serum samples it is convenient
to graph the electrical resistance (obtained at 0°C with a sensitive conductivity meter) of NaCl solutions versus the NaCl
molarities. The resistance of unknown solutions can then be
related to this standard curve and expressed as the relative
salt concentration (RSC) of the unknown [28].

1. Isotonic Veronal Buffer (Buffer 1)

Buffer stock for this buffer is made by weighing 83.0 g of
NaCl and 10.19 g of Na-5,5 diethyl barbiturate into a 2-liter
volumetric flask. Water is added to about 1.5 liters and the
pH is adjusted to 7.35 $\pm$ 0.05 with 1 N HCl. The final volume
is brought to exactly 2 liters [32].

For use the buffer concentrate is diluted 5-fold with distilled water in which 0.1% gelatin (final concentration) has
been dissolved. The gelatin is generally added to all the buffers used in hemolytic assays to prevent protein adsorption and
nonspecific lysis. The gelatin is dissolved by gentle heating
in a small aliquot of the distilled water used to dilute the
stock solution.

2. Me^{++} Isotonic Veronal Buffer (Buffer 2)

Buffer 2 is buffer 1 to which has been added a stock solution of $CaCl_2$ (0.15 M) and $MgCl_2$ (1.0 M). Of this solution, 1
ml is added to each liter of buffer 1. The final metal concentration is 0.15 mM Ca^{++} and 1 mM Mg^{++}.

3. Low-Ionic-Strength Buffer with Me^{++} (Buffer 3)

Buffer 3 is made by first mixing four parts of buffer 1
with six parts of 5% dextrose in distilled water. Then

1 ml/liter of the Ca^{++} and Mg^{++} stock solution is added as for

buffer 2. Others alter this method by mixing equal volumes of

buffer 1 and the dextrose solution, thereby increasing the RSC.

Alternatively, this buffer can be made with sucrose or mannitol

instead of dextrose [27,33].

4. EDTA Buffer

Another buffer is prepared by mixing one part of a 0.1 M

EDTA stock at pH 7.4 (conveniently prepared by titrating disodium

ethylene-diamine tetraacetic acid of the highest grade with 2 N

NaOH) with nine parts of buffer 1.

B. Preparation of Cellular Intermediates

Sheep cells (E) are generally used for the study of human

and guinea pig complement. After they are washed in buffer 2,

then in EDTA buffer for removal of the serum, the E are stan-

dardized spectrophotometrically by lysing of a precisely mea-

sured aliquot in water, and reading of the absorbance at 541 or

412 nm (convenient absorbance wavelengths for hemoglobin [32]).

The unlysed suspension is adjusted with buffer to a convenient

absorbance, generally a density of 1-5 x 10^8 cells/ml. The

number of cells can be related to the absorbance (A) values by

use of a Coulter counter. The E are sensitized by addition of

an equal volume of an optimal amount of whole antisera or puri-

fied antibody in EDTA buffer. After a 10-min incubation the EA
are washed to eliminate components of the antiserum that are not
bound, then suspended in a convenient buffer, usually buffer 2
or 3. They may be used immediately as sensitized cells (EA) or
stored for up to 4 weeks at 0°C in buffer 3 made with dextrose.
The EA are useful for C4 titrations and over-all complement
titrations (CH50 determinations) in whole serum. The discovery
of other cellular intermediates was an important breakthrough
in titrating complement components [32]. Generally, a cellular
intermediate can be formed by reacting EA successively with
purified C components, but there are convenient methods that
manipulate time and temperature and use partially purified reagents or whole serum and avoid the use of highly purified components. This is particularly true for $EAC\overline{14}$ [32] and EAC1-7 cells
[34]. The EA, $EAC\overline{14}$, EAC4, and EAC1-7 are highly stable and can
be kept for a few weeks at 0°C. Functionally pure components
can be obtained commercially, as can certain cellular interme-
diates (see Section III). Unstable $EAC\overline{142}$, $EAC\overline{1423}$, $EAC\overline{14235}$,
EAC1-6, EAC1-8, and EAC1-9 should be used immediately after
they are prepared. Stable $EAC\overline{142}^{ox}$ are made with human C2, oxi-
dized with iodine [35]. Cellular intermediates preferably
should carry a sufficient excess of cell-bound C components in
order that reactivity with limiting dilutions of the component
to be titrated will be rapid and maximal.

C. Choice of Antibody

For the routine titration of complement activity in serum, E can be sensitized with whole antiserum from rabbits, that has been produced by immunization of rabbits with erythrocyte stroma [32]. The antibodies produced by this method are a mixture of 19S and 7S classes, but are primarily IgM. The antisera should be heated at 56°C for 30 min to prevent lysis by the complement in the antisera. For certain applications, such as immune adherence [36], phagocytosis enhancement [37], and the Cl titration [38], it may be important to have a single antibody class on the cells. Certain _in vivo_ studies of immune hemolysis show a distinctly different behavior of sensitized cells, depending upon the antibody used to sensitize the cells [39]. The IgM antibody (generally the most useful antibody for _in vitro_ studies of complement components) can be purified by ammonium sulfate precipitation, followed by Sephadex G-200 gel filtration [38].

The study of immunoglobulin structure has been sufficiently advanced so that binding configurations and molecular sites where certain complement component interactions occur are partially known. Early studies by Ishizaka [11] revealed that aggregation of certain IgG immunoglobulins was sufficient to bypass a requirement for antigen. Moreover, the aggregated Fc portion was fully active [40]. Recent studies by Allan and Isliker [41] strongly implicate a small peptide containing a single trypto-

phan residue near the hinge in the Cl binding site. Not all

immunoglobulins activate and fix Cl; others, such as IgA and

IgE, are capable of activating an alternate pathway to C3

cleavage [42]. The $F(ab')2^{gp}$ may have binding sites for compo-

nents of the alternate pathway [43]. Recently, the pentameric

Fc subunit of human IgM and its monomer have been shown to be

fully active in fixing human complement [44].

D. CH50 Assay

The CH50 assay [fully described in Ref. 32] measures the

total hemolytic activity in serum that produces 50% lysis.

To determine the resultant hemolytic activity of all nine

components as they occur in serum, dilutions of whole serum are

added to standardized EA under precisely controlled conditions

of ionic strength, pH, and optimal Ca^{++} and Mg^{++} concentration.

Incubation is at 37°C for 60-90 min. Lysis is determined spec-

trophotometrically. A sigmoidal dose-response curve is genera-

ted when the percentage of cells lysed is plotted against the

serum concentration. The amount of serum that produces 50%

lysis can be determined from a linear transformation of this

curve on logarithmic paper. This amount is said to contain one

CH50 unit. The actual number of units (the CH50 titer) in any

serum is a function of the assay conditions chosen. Several reproducible and well-standardized methods are available. The CH50 determinations have been very useful for diagnosing and following the activity of certain human diseases [45].

E. C1 Titration

The study of C1 and its subcomponents has led to increased understanding of how the complement system is activated by certain immune complexes. The binding of C1 to antibody is mediated through the C1q subcomponent, which is multivalent for antibody. C1q can precipitate aggregated immunoglobulins [46] or antigen-antibody complexes from solution; this property has applicability in the diagnosis and understanding of human diseases [47]. C1q is labile and may be purified by rapid precipitation [48,49]. In the presence of calcium ions, it combines with C1r [50] and C1s. It is of interest that the addition of EDTA dissociates the molecule, but still allows C1q to interact with immunoglobulins. It is possible to use radio-labeled $\overline{C1s}$ in an immunoassay to quantitate the amount of $\overline{C1s}$ that can be dissociated by EDTA from antigen-antibody complexes after they have reacted with $\overline{C1}$ [51]. This method represents a modification of the $\overline{C1}$ fixation and transfer reaction described by Borsos and Rapp [52]. The $\overline{C1}$ fixation and transfer is used to quantitate particulate-bound antibody that combines with $\overline{C1}$. The $\overline{C1}$ fixation and transfer test has as its basis the quantitative assay of $\overline{CI}$ that dissociates from immune complexes and transfers to receptors on EAC4 cells for which it has a greater affinity. The number of C1 molecules that transfer can then be quantitated by procedures identical with the C1 hemolytic assay described in the following procedure.

(i) The C1 assay, like most component assays, has undergone an evolution. The earliest studies by Hoffman [53] were a major advance, but the physiologic ionic strength allowed C1 to dissociate from antibody, thus decreasing the titration efficiency. The optimal low ionic strength and the convenient use of EAC4 were described by Borsos and Rapp [27]. The suitability of IgM as a cell sensitizer and an estimate of the molecular titration efficiency were later additions [22]. Tests to distinguish the proenzyme form of C1 from $\overline{CI}$, and $\overline{CIs}$ from $\overline{CI}$ were described more recently [54,55]. With this C1 assay, it is convenient to titrate solutions containing as few as 10^8 C1 molecules per milliliter.

(ii) The intermediate used to titrate C1 is the EAC4 cell. These are usually stored at 1×10^9 cells/ml, and are prepared from $EAC\overline{I}4$ [32] by incubation of $EAC\overline{I}4$ with 0.01 M EDTA (add 0.11 ml of 0.1 M EDTA stock to each ml of $EAC\overline{I}4$) for 10 min at 37°C, followed by two original volume washes in EDTA buffer. An alternative method has been described and is useful [56]. The cells are carefully suspended in the low-ionic-strength buffer described previously. The EDTA washing procedure removes all the Cl activity from $EAC\overline{1}4$ by dissociation of Cl into its subcomponents Clq, Clr, and Cls [57], which are washed away.

The EAC4 are stable for two to four weeks when stored at 0°C; just before a titration they are standardized to a density of 1.54×10^8 cells/ml by lysis of an aliquot of the cell suspension (usually 0.5 ml is lysed with 7.0 ml of water). The absorbance at 412 nm is determined, and from this value the cell suspension is adjusted with buffer 3 so that when a 0.5 ml aliquot is lysed in 7.0 ml of water, an absorbance of 0.930 is obtained.

(iii) The 0.5 ml aliquots of the standardized EAC4 are dispensed in tubes. Then 0.5 ml of appropriate Cl dilutions in the same buffer are added at timed intervals, mixed, and incubated. After each tube has been incubated at 30°C for 10 min, 0.5 ml of C2 containing 50-100 molecules/cell is added, and the resultant mixture is incubated for an additional 10 min. The final step is the addition of 6 ml of $C^{gp}EDTA$, which has been prepared by dilution of whole guinea pig serum to 1/50 in 0.01 M EDTA buffer at 0°C.

The reaction tubes are placed at 37°C and lysis is allowed to reach an end point in 90 min, with occasional shaking. The tubes are then centrifuged at 1000 x $\underline{g}$ for 5-10 min to remove the cells from the fluid, and the absorbance (A) of the hemoglobin in the supernate is obtained spectrophotometrically. A com-

plete lysis control (100% lysis) and a cell blank (cells plus
buffer) are prepared and taken through all the incubation steps.
The EAC4 cells are incubated with C2 and C^{gp} EDTA to ensure that
no Cl has been left on the cells or is in any of the other reagents. This control tube is termed a CBC (cell blank plus complement) and the lysis in this tube should not be greater than
5%. Any greater degree of lysis would suggest that the reagents
are contaminated with Cl. Generally, duplicate tubes are prepared.

(iv) The percentage of lysis in each tube is determined
from the following equation:

$$\underline{y} = \% \text{ cells lysed} = \frac{A \text{ (sample)} - A \text{ (CBC)}}{A \text{ (100\%)} - A \text{ (cell blank)}} \text{ X } 100.$$

From the $-\ln(1 - \underline{y})$, the average number of Cl molecules per
cell (conveniently called $\underline{z}$) for each dilution is obtained.
This mathematical relationship is derived from the Poisson distribution [see 31].

The $-\ln(1 - \underline{y})$ when $\underline{y} = 0.63$ (63% lysis) is 1.00, which
means there is an average of one effective molecule per cell.
As only 63% of the cells hemolyze, it is obvious that some cells
have no effective molecules and others have more than one. For
instance, to obtain 95% lysis an average of three effective
molecules per cell is required. To obtain the total number of
molecules in a preparation, the dilution that gives a $\underline{z}$ value of
1.00 is determined from the titration graph and multiplied by
the number of cells. The result is usually expressed on a

molecules/milliliter basis. The linear dose-response for a
typical titration is shown in Fig. 2. Deviations from linearity
are found for Cl that is not fully active or for the precursor
form of Cl in whole serum [33]. That the number of molecules
determined hemolytically is closely equivalent to the number of
Cl molecules measured by chemical means has been proven by quantitative protein assays on highly purified samples [58]. The
validity of the titration theory has been further indicated by
the use of known amounts of purified IgM from antiserum prepared
against sheep cells [59].

III. PURIFICATION OF COMPLEMENT COMPONENTS
AND OTHER PROTEINS THAT INTERACT WITH THE COMPLEMENT SYSTEM

Isolation of C3 by protein separation methods and purification of C2 by a combination of nonspecific and specific affinity
methods will be described as examples of the mode of purification of complement components. (For references covering the
purification of other proteins, see Table 1.)

A. Purification of C3

(i) In order to assay C3 in various stages of purifica-
tion [60], $EAC\overline{14}\,\overline{2}$ cells (prepared from $EAC\overline{14}$ and C2) and C5-C9
are used [28,61]. (All of these reagents can be prepared or

they can be purchased from the Cordis Co., Miami, Florida; the

methods of assay are found in Ref. 61.) Other materials needed

are DEAE- and carboxymethyl (CM)-cellulose (from Brown Paper

Co., Berlin, N. H.), Pevikon C-870 (from Mercer Chemical Corp.,

New York, N. Y.), and the ultrafiltration collodion bag appara-

tus (from Carl Schleicher and Schuell Co., Keene, N. H.). Cel-

lulose is washed and packed into columns under gravity, as

described by Nelson et al. [28]. Preparative electrophoresis

in a Pevikon block is done as described by Müller-Eberhard [62].

All fractionation procedures are performed at 3°C unless speci-

fied otherwise.

      (ii) Guinea pig serum (150-200 ml) is adjusted to pH 7.5

with 1 M acetic acid and diluted with ice-cold distilled water

until the conductivity corresponds to that of 0.04 M NaCl. The

resulting precipitate is allowed to aggregate for 30 min, then

removed by centrifugation at 1500 x g at 0°C for 45 min. The

clear supernatant is applied at a flow rate of 250 ml/hr to a

DEAE-cellulose column 5.0 x 70 cm, equilibrated with pH 7.5

starting buffer containing 5 mM phosphate, 33 mM NaCl, and 1 mM

EDTA. The column is then washed with 1500 ml of starting buf-

fer, then a linear gradient is developed by the gradual addition

(150 ml/hr) of 1500 ml of a pH 7.5 buffer containing 5 mM phos-

phate, 243 mM NaCl, and 1 mM EDTA to a mixing chamber containing

1500 ml of the starting buffer.

      (iii) About 400 ml of eluate containing C3 is pooled, concentrated to 10 ml by ultrafiltration, and diluted to 20 ml with

an appropriate NaCl solution to adjust the conductivity to that

of 0.08 M NaCl. The pH is then brought to 5.0 with 1 M acetic

acid. The material is applied at a rate of 20 ml/hr to a CM-

cellulose column 2.5 x 40 cm, equilibrated with pH 5 starting

buffer containing 0.02 M acetate and 0.07 M NaCl. The column

is washed with 150 ml of starting buffer. A linear gradient is

then developed by the continuous addition (75 ml/hr) of 450 ml

of pH 5 buffer containing 0.02 M acetate and 0.19 M NaCl to a

mixing chamber containing 450 ml of starting buffer. Fractions

from the cellulose column are collected in tubes containing an

appropriate amount of 0.5 M Tris·HCl to bring the pH to 7.5. A

volumetric collecting device is used, since its rapid discharge

of each fraction promotes mixing with the Tris buffer.

(iv) The eluate containing C3 is concentrated to 1 ml by

ultrafiltration and dialyzed for 4 hr against 500 ml of Tris

buffer (pH 8.6), ionic strength 0.05. The material is applied,

12 cm from the cathode, to a Pevikon block (1 x 10 x 50 cm)

equilibrated with the pH 8.6 Tris buffer. Electrophoresis is

for 48 hr at 400 V, with a current of 20 mA and a gradient of

2.5 V/cm of Pevikon block. After electrophoresis, the block is

sliced into 1 x 1 x 10 cm segments, and each segment is eluted

four times with 10 ml of pH 7.5 buffer containing 5 mM phosphate,

0.15 M NaCl, and 1 mM EDTA. Only the first fraction of each

eluate is analyzed for C3 activity, but all four fractions are

pooled if the first eluate contains C3.

(v) A pool of about 300-500 ml of the electrophoretically
fractionated C3 is concentrated to 30 ml and diluted with 5 mM
phosphate buffer (pH 7.5) until the conductivity corresponds to
that of 35 mM NaCl. The pH is adjusted to 7.5 if necessary.
The material is applied at a rate of 15 ml/hr to a DEAE-cellulose
column (2 x 30 cm) equilibrated with pH 7.5 starting buffer, containing 5 mM phosphate, 28 mM NaCl, and 1 mM EDTA. The column
is washed with 100 ml of this starting buffer, then a linear
gradient is developed by the addition (30 ml/hr) of 250 ml of
pH 7.5 buffer, containing 5 mM phosphate, 193 mM NaCl, and 1 mM
EDTA, to a mixing chamber containing 250 ml of starting buffer.

(vi) The eluate containing C3 can be further processed
through the Pevikon-block electrophoresis and CM-cellulose chromatography steps, with a slight gain in purity. These steps
are necessary for certain applications, particularly if there
are significant amounts of other C components. These procedures
are similar to those described in the preceding steps, except
that they are smaller in scale. The purity of the final product
is determined by immunoelectrophoresis and analytical disk
acrylamide electrophoresis, and by functional titration of all
the other components. The specific activity should be 80- to
150-fold that of serum.

B. Purification of C2 by Specific Affinity to EAC4

The purification procedure of C2 described here is modified
from that of Mayer et al. [23]. Partially purified guinea-pig
C2 is made according to Nelson et al. [28] or Borsos et al. [63].
The EAC4 cells are prepared according to Borsos and Rapp [27].
About 235 ml of EAC4 (1 x 10^9 cells/ml) are sedimented by centrifugation in a 250-ml centrifuge bottle. The supernatant
fluid is removed and the sedimented cells are warmed to 30°C.
A partially purified C2 preparation (about 5 x 10^{11} SFU/ml) is
diluted 1/5 in buffer A, pH 8.5, ionic strength 0.0544, containing

199 mM sucrose,	4.94 mM sodium barbital,	42.6 mM NaCl,
2 mM $MgCl_2$,	0.3 mM $CaCl_2$,	0.1% gelatin;
(HCl was used to adjust the pH)		

and warmed to 30°C; then 200 ml of it is thoroughly mixed with
the cells. After the cell suspension is incubated for 10 min
at 30°C, the resulting EAC42 are sedimented by centrifugation
at room temperature (24°C) and washed twice with 200 ml of buffer
B, pH 8.5, ionic strength 0.0686, buffer containing

170 mM sucrose,	4.94 mM sodium barbital,	56.8 mM NaCl,
2 mM $MgCl_2$,	3 mM $CaCl_2$,	0.1% gelatin;
the pH was adjusted with HCl.		

After removal of the second wash fluid, C2 is eluted from
the cells with 75 ml of buffer C (pH 6.5, ionic strength 0.147).

This buffer was prepared similarly to buffer A except that the pH was adjusted to 6.5 with HCl, and the gelatin concentration was reduced to 0.05% at 0°C. After 10 min at 0°, the EAC4 are removed by centrifugation at 0°C, and the supernatant, containing virtually no other serum protein except C2, is collected. The supernatant fluid is, however, contaminated with hemoglobin and other proteins that have leaked out of the erythrocytes.

If it is desirable, the erythrocyte proteins can be removed by preparative acrylamide gel disk electrophoresis (modified from the description of Davis [64]). The supernatant containing C2 is concentrated about 10 times by ultrafiltration at 5°C in the apparatus purchased from Schleicher and Schuell Co. Owing to the high gelatin concentration, the concentrated contents of the collodion bags may solidify. Therefore, the collodion bags are immersed at 37°C in the buffer used for electrophoresis to liquefy the solution, then the contents are layered on the stacking gel. After the remaining space at the top of the tube is filled with buffer, the sample is electrophoresed until the tracking dye, bromphenol blue, emerges from the anodal end. Gel segments are eluted in isotonic saline-Veronal buffer, buffer A, containing 0.01% gelatin or 2% lysine. The lysine in the C2 preparation may be removed by dialysis. Electrophoresis may be repeated to further eliminate traces of impurities.

C. Other Complement Components and Some Proteins
That Interact with Complement

The references for the purification of other complement

components are given in Table 1.

TABLE 1

Purification Methods and Their References

Complement	Guinea pig	Human
C1	65	58
C1q	----	48
C1s	----	55
C1r	----	50
C4	28	69
C2	23	70
C3	60	71
C5	61	71
C6	28	72
C7	28	----
C8	28	73
C9	66	74
C1 inhibitor	67	75
C3b inactivator	67	76
Properdin	----	77
Factor B (heat-labile serum cofactor for cobra venom)	68	42, 78

Purity should be judged by standard, sensitive protein separation tests for homogeneity, by immunochemical tests, and by
assays for other components. The absence of other detectable
components is called functional purity, and is of particularly
great importance in hemolytic or biological assays.

A natural inhibitor of C6 was originally fractionated from
rabbit serum by Nelson and Biro [79]. This inhibitor was present in both human and guinea pig serum. A potent anticomple-
mentary protein can be isolated from cobra venom by the methods
of Shin et al. [80] or Müller-Eberhard and Fjellström [81].

IV. BIOLOGICAL ASSAYS

A. Bioassay for Anaphylatoxins

The anaphylatoxin activity of C3a or C5a can be measured
by the contraction of isolated guinea pig ileum caused by anaphylatoxin. Characteristically, the contraction of the ileum
commences several seconds after the introduction of anaphylatoxin and, on repeated addition of the same anaphylatoxin preparation, the contraction of ileum diminishes in intensity, a
phenomenon called tachyphylaxis. The contraction can be inhibited by antihistamine drugs. An apparatus designed according to
Randall et al. [82] is used for the measurement of ileal contraction.

Male albino guinea pigs (400-500 g) are killed by a blow
at the base of the skull. The abdominal fur is moistened and
the lower abdominal skin is opened. The ileum is cut about
2-3 cm proximal to the ileocecal junction, and gentle continuous traction is applied with forceps to the proximal cut end.
The entire ileum and jejunum (about 3 ft long) are drawn out
until firm resistance is met at the duodenal-jejunal junction.
The intestine is cut at this point and placed in about 20 ml of
Kreb's buffer in a petri dish; several 2- to 3-ml portions of
buffer at 37°C are introduced by pipette into the proximal
jejunum and allowed to flow through under gravity. After a
cleansing, the intestine is transferred to fresh buffer in a
second petri dish. The proximal end is labeled with a 2-0 black
twist silk tie; about 1 cm of the distal end is draped over the
edge of the petri dish; a single loop is formed in one end of a
12-in. piece of silk, passed around the entire circumference of
the intestine about 2-3 mm from the distal end, pulled tight,
and "squared." The intestine is then pulled gently by the
thread until the desired length (3/4 in.) extends beyond the
edge of the dish, and is cut at the edge. The process is then
repeated. The isometric tension of a piece of smooth muscle is
only slightly dependent on length, so precise control of the
segment length is not critical. The ileum segment is tied
between a fixed steel bar and a rod attached to a pressure
transducer. The ileum segments sit in a 37°C water-jacketed

10-ml tissue bath which contains a Kreb's-type buffer. Air or
95% O_2-5% CO_2 is bubbled through the tissue bath. The buffer
consists of the following chemicals in grams/liter:

NaCl	7.175	$CaCl_2 \cdot H_2O$	0.252
KCl	0.344	$NaHCO_3$	1.0
KH_2PO_4	0.157	Dextrose	1.0
$MgSO_4 \cdot 7H_2O$	0.288	Atropine	30 μg/liter

Concentrated histamine is added into the chamber with a
syringe, in a small volume so as to minimize changes in the buffer. After each contraction of the muscle, the chamber is emp-
tied and refilled with fresh solution. The period between suc-
cessive additions of histamine should be kept uniform (about
1 min) in order to obtain optimal equilibration. A properly
equilibrated piece of intestine generally responds to a final
concentration of histamine·HCl of about 5 ng/ml of tissue-bath
fluid.

When equilibration has been attained, the test materials
may be added and contraction of the muscle is recorded over a
period of several minutes.

B. Assay for Leukocyte Chemotaxis

The following procedure is from Boyden [83] as modified by
Snyderman et al. [84]. The techniques used to assay leukocyte
chemotaxis are similar for polymorphonuclear leukocytes (PMNs),

macrophages, eosinophils, and lymphocytes. Only the technique

for the measurement of chemotaxis of human and rabbit PMNs is

described. For other cell types, appropriate references are

given at the end of the section.

The modified Boyden chambers used for the chemotaxis assay

can be purchased from Neuroprobe Inc. 7400 Arden Road, Bethesda,

Md., or from Ahlco Machine Company, 232 Arch Street, New Briton,

Conn. The most satisfactory cellulose ester filters for rabbit

neutrophil chemotaxis can be obtained from the Sartorious Divi-

sion, Brinkman Instruments, Inc., Westbury, N. Y., or from

Schleicher and Schuell, Keene, N. H. Filters of 1.2 μm pore

size are desirable. For chemotaxis of human and mouse peri-

pheral blood neutrophils, filters with 5.0 μm pore size (Milli-

pore Filter Corp. Bedford, Mass.) are optimal.

Rabbit neutrophils are obtained by injection of about 150 ml

of 0.1% (w/v) shellfish glycogen intraperitoneally into normal

rabbits. About 6 hr after the injection of glycogen, the rab-

bits are tied to a restraining board. A 15-gauge needle with

holes bored in its sides is inserted into the peritoneal cavity

and the cavity is lavaged with heparinized saline (2 units/ml).

In most cases, rinsing of the peritoneal cavity with 150 ml of

heparinized saline is sufficient to obtain at least 5×10^8

leukocytes (>95% PMNs). The peritoneal fluid is collected by

kneading the rabbit's abdomen and allowing the fluid to drain

from the needle into 50-ml plastic centrifuge tubes. The fluid

is then centrifuged at 4°C for 10 min at about 500 x $\underline{g}$, the
supernatant is discarded, and the cells are suspended at a density of 2.2 x 10^7 cells/ml in Gey's balanced salt solution containing 2% bovine serum albumin (Gey's medium) [84].

Human PMNs are obtained by drawing venous blood into a
heparinized plastic syringe. The final concentration of heparin
should be about 20 units/ml of blood. The blood is then diluted
with an equal volume of 3% (w/v) high-molecular-weight dextran
(Pharmacia Fine Chemicals, Piscataway, N. J.) in saline. The
erythrocytes are allowed to sediment for about 30 min at room
temperature, and the leukocyte-rich supernatant is withdrawn.
The supernatant is centrifuged (500 x $\underline{g}$ for 10 min) and the
resultant supernatant is discarded. The erythrocytes contaminating the leukocytes are removed by hypotonic lysis: the cells
are mixed with 10 ml of 0.2% NaCl for 20 sec then 10 ml of
1.6% NaCl is added. The cells are centrifuged as before, washed
once in Gey's medium, then suspended in Gey's medium to a final
density of 2.8 x 10^6 cells/ml. In most cases, the leukocytes
will contain about 70% PMNs.

Samples tested for chemotactic activity are diluted in
Gey's medium and introduced into the lower compartment of the
fully assembled chamber, with a filter in place. The chamber
should be tilted forward to prevent the accumulation of air bubbles under the filter. When the filter becomes wet, the chamber
is placed on a flat surface and the cell suspension is introduced

into the upper chamber, while the chemotactic sample is being
continuously added to the lower chamber to keep the fluid level
in both compartments equal. After a 3-hr incubation at 37°C
(in humidified air), the fluid is aspirated from both compartments of the chamber.

Filters are taken out with a fine forceps. For the staining of several filters simultaneously, a cage with compartments
that carry a single filter made of stainless-steel wire mesh or
paper clips attached in series to a wire is convenient.

The filters are rinsed gently in normal saline, treated
with 95% ethanol for 10 sec, rinsed in distilled water, stained
5 min in Mayer's hemotoxylin, rinsed in distilled water, then
in 1% acid alcohol for 1 min, rinsed in distilled water, then
in blueing agent (2 g of $NaHCO_3$ and 20 g of $MgSO_4$ in 1 liter of
distilled water), for 2 min rinsed in distilled water, then in
70% ethanol for 1 min, rinsed in 95% ethanol for 1 min, in
absolute ethanol for 1 min, then finally cleared in xylene.
The membrane is mounted in xylene on a microscope slide with
the lower side of the filter (in reference to its position in
the chemotaxis chamber) facing upward.

Ten randomly chosen fields are counted under a high-dry
objective (X400) and the average count per field is used to
express the chemotactic activity of the sample. A microgrid,
which can be inserted into the eyepiece of most microscopes,
aids in making the counting procedure more precise.

Methods for the assay of chemotactic activity by use of various leukocytes are given in the references in Table 2.

TABLE 2

Chemotactic Methods and Their References

Source	Neutrophil	Eosinophil	Macrophage	Lymphocyte
Human	85	88	90	--
Rabbit	83	--	91	--
Guinea pig	86	89	86	--
Rat	--	--	--	92
Mouse	87	--	--	--

C.   Immune Adherence

Immune adherence [36] depends upon a receptor site on C3 for mammalian cells, or rabbit or guinea pig platelets. The C3 can be bound to antigen-antibody complexes as a result of the normal activation of the complement sequence [36], or chemically by use of tanned erythrocytes and purified C3. The method can be adapted for titration of antigen, antibody, or C3. It is also a useful way to determine if C3 has reacted with an antigen-antibody complex; however, the C3 inactivator may have to be inhibited if whole serum is used [10].

Immune complexes carrying activated C3 are diluted serially, twofold, in 0.01 M EDTA buffer. The diluted complexes are thoroughly mixed with 0.1 ml of washed type-0, Rh-positive human

erythrocytes (indicator cells), 1 x 10^8 cells/ml in EDTA buffer.

After 60 min of undisturbed incubation at 37°C, the agglutina-

tion pattern is read. The end point is considered to be the

dilution of antigen, antibody, or C3 that causes the agglutina-

tion of 50% of the human indicator erythrocytes. The end points

can be judged by a 2+ agglutination pattern or microscopically.

Controls containing only indicator cells, dilutions of antigen

without antibody, and of antibody without antigen are needed.

For routine titrations of antigen or antibody, prior tests are

necessary to determine the optimal dilutions of antigen or anti-

body.

D. Phagocytosis

The enhancement of phagocytic activity by activation of

complement appears to be a function of two components, C3 and

C5. The activation of these two components so that they will

enhance phagocytosis has been studied primarily in three sys-

tems: (1) enhancement of phagocytosis of erythrocytes that have

been sensitized (EA) and have reacted with various complement

components in a manner analogous to the formation of cellular

intermediates [37]; (2) enhancement of phagocytosis of pneumo-

cocci sensitized with specific antibody [93]; and (3) phagocy-

tosis of pneumococci sensitized by exposure to the naturally

occurring immunoglobulins in nonimmune serum [94]. The important issue of whether the latter is directed specifically at
the pneumococcus, or is a cross-reacting antibody of low affinity, or is a nonspecifically bound immunoglobulin has not been
settled. All these systems lend themselves to a quantizative
appraisal of the phagocytic process, but since the erythrocyte
system has outstanding simplicity, and since much is known about
the quantitation of C fragments bound to this particular antigen, the erythrophagocytic method developed by Gigli and Nelson,
Jr. [37] is described.

Erythrocytes sensitized by the method described above to
make EA (using specific classes and optimal concentrations of
antibody when this is the important issue) are reacted sequentially with purified complement components. The amount of any
component should be carefully considered, and it is preferable
to use complement components from the species that supplies the
cells. We do not know of any detailed study of the dose-response
nature of the phagocyte response to various bound components.
Erythrocytes (5 x 10^7 ml) were exposed to 4 x 10^7 polymorphonuclear leukocytes (see Section IV,B for collection methods) in a
siliconized flask with gentle agitation at 37°C, in buffer, for
30-45 min. The total reaction volume was 5 ml. Samples (2 ml)
are removed and added to 5 ml of 5 mM phosphate buffer, pH 7.5.
The noningested erythrocytes lyse; the mixture is then centri-

fuged at 0°C to remove the leukocytes. The number of nonphago-
cytosed erythrocytes that is lysed is determined from the absorbance of the supernatant hemoglobin. This value is subtracted
from the original number of cells present, after an appropriate
correction is made for any absorbance in the control tubes containing only leukocytes. The difference represents the number
ingested. The optimal buffer for rapid phagocytosis has a RSC
of 0.1 (less than physiologic), and the buffer used is a Hanks
balanced salt solution containing 0.2% gelatin.

The results of Gigli and Nelson's original study are shown
in Table 3. Increased phagocytosis occurs after the addition

TABLE 3

Erythrophagocytosis [modified from Ref. 37]

Cellular intermediate	No. of ingested cells x 10^6
E	5.0
EA	6.9
EAC1	8.7
EAC14	8.0
EAC142	8.4
EAC1423	38.1
EAC14235	38.6
EAC142356	40.2
EAC1423567	41.9
EAC14235678	35.0

of C3 to the $EAC\overline{142}$ intermediate, and presumably depends on a
receptor for C3 on the PMNs. It is probably analogous to the
immune adherence receptor on erythrocytes.

V. ULTRASTRUCTURE OF Clq AND LESIONS IN MEMBRANES PRODUCED BY
    THE INTERACTION OF ANTIBODY AND C: SPECIMEN
    PREPARATION AND INTERPRETATION OF RESULTS

Elucidation of the structure of Clq and characterization
of the membrane lesions produced by antibody and C have relied
principally upon one preparative technique, negative staining.
Analysis of sectioned membranes has been disappointing in its
yield of information. Humphrey et al. [95] fixed, embedded,
and sectioned human erythrocytes that had been lysed with water
or by interaction with Forssman antibody and C. The erythrocyte
unit-membrane structure was preserved after lysis in water, while
membranes of cells subjected to complement lysis were thickened
and foamy, but no structures resembling holes could be identi-
fied in the latter.

The materials and techniques used in negative staining of
macromolecules and membranes are described in the book edited
by Kay [96]. In it, D. E. Bradley outlines several methods for
the preparation of support films and R. W. Horne supplies protocols for the negative staining of cells, cell organelles, and
macromolecules, and discusses the characteristics of the various
heavy-metal salts used as negative stains. (Materials for specimen preparation may be purchased from E. F. Fullam, Inc.,
P. O. Box 444, Schnedectady, N. Y. 12301, or Ladd Research Industries, Inc., P. O. Box 901, Burlington, Vermont, 05401.) The
following discussion presupposes a general knowledge of the

preparation of support films and negative staining and emphasizes those areas where attention to detail will most likely yield successful results.

Negative staining, in principle, is exceedingly simple. It consists of suspending material to be stained in a solution of a heavy-metal salt and drying it on a carbon film supported on an electron microscope grid. The suspended material protrudes above the dry film and irregularities in its surface trap small amounts of the metal salt. Electrons pass through the material to be stained, but are almost completely back-scattered by the heavy-metal salt. The material and some of its contours are seen in negative contrast on a photographic print. In practice, good negative stains are achieved only with properly prepared support films, clean staining solutions, and an optimum concentration of protein. It becomes more and more difficult to prepare good negative stains as the size (molecular weight) of the protein decreases. Thus, bacteria, large viruses, and cell membranes are relatively easy to stain, simply because they trap the solution of heavy-metal salts quite efficiently, thus creating the contrast essential for visualizing the structure. Soluble proteins are much more difficult to stain and the dimensions and substructure of proteins below a molecular weight of 100,000 should be interpreted with extreme caution. Mellma et al. [97] studied the effect of uranyl salts upon the low-molecular-weight proteins pepsin, β chymotrypsin, trypsin,

and papain and found that there was no irreversible loss of
enzymatic activity or change in hydrodynamic properties of the
enzymes after treatment with these salts, provided that the pH
of the solutions did not precipitate the protein. They conclu-
ded, however, that only general shape and average size of these
small proteins could be estimated. Horne [in Ref. 96] found
that viruses were infectious after suspension in solutions of
uranium and tungsten salts. These experiments indicate that
negative staining does not necessarily destroy the integrity of
the material being stained; nevertheless, it is advisable to
use more than one pH and more than one heavy-metal salt when
negative staining an unknown protein.

A.   Support Films

Clean, hydrophilic, and stable carbon films are the sine
qua non of high-quality negative stains. Valentine et al. [98]
and Dourmashkin et al. [99] achieved this by evaporating carbon
onto freshly cleaved mica, stripping the carbon film on the surface of a buffered solution of the molecules to be examined,
placing grids on the film, and picking them up with a square of
newspaper. A more convenient method is to prepare carbon-coated
nitrocellulose membranes supported on copper grids ahead of the
actual staining procedure. Such grids may be stored without
deterioration for months in an evacuated, clean, vacuum desic-

cator. For high-resolution work, grain in the carbon film
should be minimized by evaporation at low amperage for 5 to 10
min with the grids placed as far from the carbon source as possible, at least 10 cm. After the film has been evaporated, a
most important step is to render it hydrophilic by cleaning it
in a glow discharge to remove the film of oil that is deposited
by all oil-pumped vacuum evaporators. In setting up the evaporator, the grids should be arranged close to one of the uninsulated feed-through posts on which the glow discharge will be
generated. The pressure in the chamber is raised to between
300 and 500 μm of mercury by the admission of air, and a high-
frequency induction coil (Vacuum leak detector, Cat. No. 9675-
L10, A. H. Thomas Co., P. O. Box 779, Philadelphia, Pa. 19105)
is touched and held for 4 min on the uninsulated post or its
connector, where it emerges on the underside of the baseplate
of the evaporator. If the room light is dimmed, the glow discharge around the post can be monitored for intensity, which
will vary according to the amount of oxygen admitted to the bell
jar. After 4 min the grids will be clean and, if properly
stored, will remain hydrophilic for months.

B. Stains

The most commonly used negative stains are 1% or 2% solutions of uranyl acetate, sodium phosphotungstate, and sodium

silicotungstate (the latter is available from TAAB Laboratories,
52 Kidmore End Road, Emmer Green, Reading, England). A solution
of uranyl acetate has a pH of 4.2, and it is not useful to raise
the pH since the salt begins to precipitate at pH 5.0. Solutions of sodium phosphotungstate are prepared by neutralization
of solutions of phosphotungstic acid to the desired pH with 1 N
NaOH or KOH; sodium silicotungstate solutions are neutral. These
solutions may be kept for months, but should be filtered through
a fine cellulose membrane filter (pore size 0.22 μm) before each
use.

C. Optimum Concentration of Substances to Be Stained

It is the tendency of the novice to use solutions that are
much too concentrated; the quality of a negative stain is almost
always improved by reducing, rather than increasing, the protein
concentration. Large objects such as cell membranes are compa-
ratively easy to stain; they are easy to recognize and the contrast is good. Negative staining of small proteins is much more
difficult to achieve. If solutions are too concentrated or too
dilute, they will look exactly the same in the electron microscope; both will appear to be unstained, the former because the
layer of protein is too thick to admit the heavy-metal salt that
provides the contrast and the latter because a certain minimum
amount of protein is required for the formation of an even film

of stain. Good negative stains of molecules of molecular weight

10^5-10^6 can be obtained from solutions containing between 10 and

50 µg of protein per milliliter. Below 10 µg of protein/ml,

staining becomes increasingly difficult, and yet, as was the

case with the Clq molecule, it is sometimes essential to go to

lower concentrations in order to separate the individual mole-

cules on the grid. Dourmashkin et al. [99] and Shelton et al.

[100] arrived independently at a solution to this problem; they

raised the total protein concentration by the addition to the

solution of a well-characterized protein, such as adenovirus or

glutamine synthetase, thus permitting the concentration of the

unknown protein to be reduced to 1-5 µg/ml. Two final notes of

caution are given: (1) proteins stick to surfaces, especially

to glass, and when diluting proteins for negative staining it

is well to use plastic containers and to use the diluted mater-

ial immediately; (2) proteins that have been lyophilized do not

rehydrate to give satisfactory negative stains.

D. Preparation of Membrane Lesions

Lesions may be produced by various means [101]; those pro-

duced by Forssman antibody and guinea pig C interacting with

sheep erythrocytes can be prepared as follows [see Ref. 102]:

sheep erythrocytes drawn under sterile conditions are stored in

Alsever's solution for at least one week. Fresh guinea pig

serum, adsorbed twice with packed sheep cells and titrated, is
used as a source of complement. The cells are sensitized as
described in Section II,C.

1. Procedure

    Erythrocytes are suspended in buffer 2 at a concentration
of 5×10^8 cells/ml. Appropriate dilutions of antibody and complement that produce about 100% hemolysis are used; then 0.5 ml
of antibody diluted in buffer 2 is added to 0.5 ml of cells,
and mixed vigorously during addition to ensure even distribution
of the antibody on the cell membranes. After incubation at room
temperature for 30 min with occasional mixing, 1 ml of the appropriate dilution of complement is added. The mixture is incubated
1 hr at 37°C with occasional mixing, then 0.5 ml of the mixture
is added to 3.0 ml of ice-cold buffer 2 and centrifuged for
3 min at 1500 x $\underline{g}$. The extent of hemolysis is determined spectrophotometrically [32]. The remaining 1.5 ml of sample is
mixed with 10.5 ml of ice-cold water. The lysed erythrocytes
are centrifuged at 120,000 x $\underline{g}$ and 4°C for 30 min. The pellet
is suspended in 12 ml of 0.02 M phosphate buffer, pH 7.4, and
again sedimented at 120,000 x $\underline{g}$ for 30 min. Finally, the pellet is suspended in a small amount of phosphate buffer and
negative stained.

E. Preparation of C1q

Freshly drawn human blood (200 ml) is allowed to clot at room temperature for 90 min. The serum is separated by a preliminary centrifugation at 500 x g for 20 min, followed by a second centrifugation at 900 x g for 20 min. The serum (63 ml) is centrifuged at 41,000 x g for 90 min to remove free lipid and insoluble aggregates. The clear supernatant (57 ml) is dialyzed at 4° with stirring, against 1 liter of 0.03-M ethyleneglycol Bis(aminoethyl)-tetraacetic acid (EGTA), with a relative salt concentration (RSC) of 0.036, pH 7.5, for 4.5 hr, then overnight against 1 liter of fresh EGTA. The resulting precipitate is centrifuged and washed twice with 0.03-M EGTA (pH 7.5, RSC 0.036). The washed precipitate is dissolved in 11 ml of 0.75 M NaCl in 0.01 M EDTA (ethylenediamine tetraacetic acid, pH 5.0, RSC is 0.78) and centrifuged at 41,000 x g for 20 min to remove insoluble aggregates. The supernatant is dialyzed against 1 liter of 72 mM EDTA (pH 5.0, RSC is 0.08) for 3.5 hr at 4°, and the precipitate is then washed once in the same concentration of EDTA and once in 0.07 M NaCl before it is dissolved in 7 ml of 0.75 M NaCl. This final solution of purified C1q is centrifuged at 41,000 x g for 20 min to remove insoluble aggregates and the clear supernatant is used immediately for electron microscopy. The purity of the preparation can be checked by radial immunodiffusion in individual agarose plates containing antiserum to C1q, IgG, IgM, or IgA.

F. Staining Procedure

The following technique has proved to be satisfactory for
negative-staining membranes and molecules. To a 4 x 6-cm rec-
tangle of quarter-inch thick Teflon is attached a strip of
"double stick" Scotch brand transparent tape (Scotch brand
#A-8-666, Pressure sensitive, Double-stick tape). The selection
of this tape is not trivial; it has a protective, nonadhesive
strip on one side that greatly facilitates handling and the
adhesive glue is less likely to adhere to the grid than with
other double-stick tapes. The tape is stuck precisely along
the edge of the block so that the grids, which are attached to
it by their outer rims, protrude beyond the block. The block
and tape are covered with one-half of a standard petri dish to
protect the tape and, during staining, the grids from dust. A
droplet of solution containing the material to be stained is
approached with the side of the grid. If the grid is hydrophilic,
a fraction of the droplet will snap onto the grid, filling it
with no overflow. The drop is allowed to remain on the grid for
3 min; it is then removed with the edge of a sliver of filter
paper and replaced with droplets of the staining solution, which
are then also removed with filter paper. The preparation is then
allowed to dry and can be examined in the microscope. If it
appears that the material to be stained needs to be fixed in
order that its structure be preserved, droplets of fixative (4%

formaldehyde solution, pH 6.5, for example) can be added to the
grid before the negative stain. If the materials are suspended
in a nonvolatile salt such as NaCl or KCl, it is advisable to
rinse with several droplets of the stain to dilute out the salt.
Such a precaution is not necessary with volatile salts such as
NH_4CO_3. The grid should be examined by reflected light as the
last droplet of stain is being removed by touching the filter
to the edge. An even film of stain should remain that cannot
be removed by capillarity. If the droplet of stain is removed
entirely so that the carbon film appears to be dry, the carbon
surface is hydrophobic and satisfactory negative staining will
not be achieved.

As a general rule, the preparations are not allowed to dry
until after stain has been added, but it is to be noted that
Medhurst et al. [103] concentrated bacteria and bacterial cell
walls by allowing a drop of suspension to dry on the grid before
staining. Complement lesions of the proper dimensions were
observed in these cells. A "one-step" method for staining membranes is to mix the negative stain with the diluted membrane
suspension and to allow a droplet of the mixture to dry on a
coated grid [102].

G. Interpretation of Results

Interpretation of the ultrastructure of small molecules of
unknown size and conformation can be very difficult. Certain

proteins, such as viruses and enzymes, are compact and of sufficient rigidity to present a very constant appearance when
negatively stained and to lend themselves to sophisticated
three-dimensional analysis [104]. Certain other proteins, such
as the immunoglobulins and C1q, can undergo gross conformational
changes and various forms may be observed by electron microscopy
[100]. It cannot be overemphasized that care must be taken to
negatively stain all solutions used in the preparation of an
uncharacterized molecule. Contamination of buffer solutions or
the heavy-metal salt solution itself with organic (bacteria) or
inorganic (dust) materials can produce images that must be eliminated from consideration when the structure of the molecule
itself is interpreted. Similarly, carbon films have inherent
structure and only those structures and substructures that have
information "above background" can be considered significant.

Component C1q has been particularly refractory to analysis,
and some recent interpretations of its ultrastructure have not
been convincing [105,106]. The C1q molecules isolated and
stained by the methods described above are revealed as delicate
structures consisting of a central subunit, connecting strands,
and terminal subunits. Each terminal subunit is further subdi-
vided into a large and small subunit, and the central connecting
strands link six terminal subunits by their larger subdivision
to the central subunit. The over-all dimension of the molecule
is about 35 nm [100].

It was recently reported [10] that C3 has a circular subunit of about 3 nm; the molecular weight of the subunit was
calculated to be around 11,000. If 17 subunits are assumed,
the whole molecular weight becomes 187,000, a figure that is
close to that arrived at by physical means.

The lesions produced by the interaction of antibody and C
with natural and artificial membranes appear as circular dense
spots surrounded by an electron-lucent outer rim. The internal
diameter, 8.5-9.5 nm (guinea pig C) or 10-11 nm (human C), of
the structures is surprisingly constant and is dependent upon
the source of complement rather than the nature of the membrane.
A thorough discussion of the lesions can be found in the com-
prehensive and well-illustrated review by Humphrey and Dourmashkin [101]. In spite of 10 years of effort, the exact nature of
these membrane lesions remains to be defined.

ACKNOWLEDGMENT

The assistance of Dr. Ralph Snyderman with the section on
chemotaxis is gratefully acknowledged.

REFERENCES

1. M. M. Mayer, Immunochemistry, 7, 485 (1970).
2. N. R. Cooper, M. J. Polley, and H. J. Müller-Eberhard,
in Immunological Diseases, (M. Samter, ed.), 2nd ed., Little,

Brown and Co., Boston, 1971.
    3. H. J. Müller-Eberhard, Ann. Rev. Biochem., 38, 389
(1969).
    4. R. M. Stroud, K. F. Austen, and M. M. Mayer, Immuno-
chemistry, 2, 219 (1965).
    5. H. J. Müller-Eberhard and I. H. Lepow, J. Exptl. Med.,
121, 819 (1965).
    6. V. H. Donaldson, E. Merler, F. S. Rosen, K. W. Kretsch-
mer, and I. H. Lepow, J. Lab. Clin. Med., 76, 986 (1970).
    7. H. J. Müller-Eberhard, A. P. Dalmasso, and M. A. Cal-
cott, J. Exptl. Med., 123, 33 (1966).
    8. H. S. Shin, R. Snyderman, E. Friedman, A. Mellors, and
M. M. Mayer, Science, 162, 361 (1968).
    9. V. A. Bokisch and H. J. Müller-Eberhard, J. Clin.
Invest., 49, 2427 (1970).
    10. K. Nishioka, Adv. Cancer Res., 14, 231 (1971).
    11. K. Ishizaka, Progr. Allergy, 7, 32 (1963).
    12. J. Sjoquist and G. Stalenhein, J. Immunol., 103, 467
(1969).
    13. S. Yachnin, D. Rosenblum, and D. Chatman, J. Immunol.,
93, 549 (1964).
    14. J. Cantrell, R. M. Stroud, and K. Pruitt, Diabetes,
21, 872 (1972).
    15. W. Willoughby, H. S. Shin, and R. Ford, Complement
Workshop Proceeding, La Jolla, California, 1973.
    16. H. J. Müller-Eberhard, in Progress in Immunology
(B. Amos, ed.), Academic Press, New York, 1971, pp. 553-564.
    17. I. H. Lepow, in Progress in Immunology (B. Amos, ed.),
Academic Press, New York, 1971, pp. 579-595.
    18. R. M. Stroud, M. M. Mayer, J. A. Miller, and A. T.
McKenzie, Immunochemistry, 3, 163 (1966).
    19. R. E. Spitzer, A. E. Stitzel, V. L. Pauling, N. C.
Davis, and C. D. West, J. Exptl. Med., 134, 565 (1971).
    20. S. Ruddy and K. F. Austen, Progr. Med. Genet., VII,
69 (1970).
    21. M. M. Mayer, in Ciba Found. Symp. on Complement (G. E.
W. Wolstenholme and J. Knight, eds.), Churchill, London, 1965,
p. 4.
    22. H. R. Colten, T. Borsos, and H. J. Rapp, Immunochemis-
try, 6, 461 (1969).
    23. M. M. Mayer, J. A. Miller, and H. S. Shin, J. Immunol.,
105, 327 (1970).
    24. J. K. Smith and E. L. Becker, J. Immunol., 100, 459
(1968).
    25. S. C. Kinsky, Biochem. Biophys. Acta, 265, 1 (1972).
    26. T. Borsos, H. J. Rapp, and M. M. Mayer, J. Immunol.,
87, 310 (1961).
    27. T. Borsos and H. J. Rapp, J. Immunol., 91, 851 (1963).
    28. R. A. Nelson, J. Jensen, I. Gigli, and N. Tamura,
Immunochemistry, 3, 111 (1966).

29. S. Ruddy and K. F. Austen, J. Immunol., 99, 1162 (1967).
30. S. Ruddy and K. F. Austen, Arthritis Rheumat., 13, 713 (1970).
31. H. J. Rapp and T. Borsos, Molecular Basis of Complement Action, Appleton-Century-Crofts, 1970.
32. M. M. Mayer, in Experimental Immunochemistry, 2nd ed. (E. Kabat and M. M. Mayer, eds.), Thomas, Springfield, Illinois, 1961, pp. 133-240.
33. L. G. Hoffman and N. P. Steuland, Immunochemistry, 8, 499 (1971).
34. F. A. Rommel and R. Stolfi, Immunology, 15, 469 (1968).
35. M. J. Polley, J. Immunol., 107, 1493 (1971).
36. K. Nishioka and W. D. Linscott, J. Exptl. Med., 118, 767 (1963).
37. I. Gigli and R. A. Nelson, Jr., Exptl. Cell Res., 51, 45 (1968).
38. H. R. Colten, T. Borsos, and H. J. Rapp, Immunochemistry, 6, 461 (1969).
39. A. D. Schreiber and M. M. Frank, J. Clin. Invest., 51, 575 (1972).
40. K. Ishizaka, T. Ishizaka, and J. Banovitz, J. Immunol., 94, 824 (1965).
41. R. Allan and H. Isliker, Experientia, 27, 725 (1971).
42. O. Götze and H. J. Müller-Eberhard, J. Exptl. Med., 134, 90S (1971).
43. A. L. Sandberg, B. Oliveira, and A. G. Osler, J. Immunol., 106, 282 (1971).
44. A. G. Plaut, S. Cohen, and T. B. Tomasi, Jr., Science, 176, 55 (1972).
45. P. H. Schur and K. F. Austen, Ann. Rev. Med., 19, 1 (1968).
46. H. J. Müller-Eberhard and H. G. Kunkel, Proc. Soc. Exptl. Biol. Med., 106, 291 (1961).
47. V. Agnello, R. J. Winchester, and H. G. Kunkel, Immunology, 19, 909 (1970).
48. K. Yonemasu and R. M. Stroud, J. Immunol., 106, 304 (1971).
49. J. E. Volanakis and R. M. Stroud, J. Immunologic Methods, 2, 25 (1972).
50. M. M. E. deBracco and R. M. Stroud, J. Clin. Invest., 50, 838 (1971).
51. R. M. Stroud, J. Lab. Clin. Med., 77, 645 (1971).
52. T. Borsos and H. J. Rapp, J. Immunol., 95, 559 (1965).
53. L. G. Hoffman, Ph.D. Thesis, School of Hygiene and Public Health, Johns Hopkins University, Baltimore, 1960.
54. M. Loos, T. Borsos, and H. J. Rapp, J. Immunol., 108, 683 (1972).
55. K. Nagaki and R. M. Stroud, J. Immunol., 102, 421 (1969).

56. T. Borsos and H. J. Rapp, J. Immunol., 99, 263 (1967).
57. I. H. Lepow, G. B. Naff, E. W. Todd, J. Pensky, and
C. F. Hinz, J. Exptl. Med., 117, 983 (1963).
58. H. R. Colten, T. Borsos, H. E. Bond, and H. J. Rapp,
J. Immunol., 103, 862 (1969).
59. H. J. Rapp and T. Borsos, J. Immunol., 96, 913 (1966).
60. H. S. Shin and M. M. Mayer, Biochemistry, 7, 2991
(1968).
61. C. T. Cook, H. S. Shin, M. M. Mayer, and K. Laudenslayer, J. Immunol., 106, 467 (1971).
62. H. J. Müller-Eberhard, Scand. J. Clin. Lab. Invest.,
12, 33 (1960).
63. T. Borsos, H. Rapp, and C. T. Cook, J. Immunol., 87,
330 (1961).
64. B. J. Davis, Ann. N. Y. Acad. Sci., 121, 404 (1964).
65. N. Tamura and R. A. Nelson, Jr., J. Immunol., 101,
1333 (1968).
66. N. Tamura and A. Shimada, Immunology, 20, 415 (1971).
67. N. Tamura and R. A. Nelson, Jr., J. Immunol., 99,
582 (1967).
68. V. Brade, C. T. Cook, H. S. Shin, and M. M. Mayer,
J. Immunol., 109, 1174 (1972).
69. H. J. Müller-Eberhard and C. E. Biro, J. Exptl. Med.,
118, 447 (1963).
70. M. Polley and H. J. Müller-Eberhard, J. Exptl. Med.,
128, 533 (1968).
71. U. R. Nilsson and H. J. Müller-Eberhard, J. Exptl.
Med., 122, 277 (1965).
72. C. M. Arroyave and H. J. Müller-Eberhard, Immunochemistry, 8, 995 (1971).
73. J. A. Manni and H. J. Müller-Eberhard, J. Exptl. Med.,
130, 1145 (1969).
74. R. Hadding and H. J. Müller-Eberhard, Immunology, 16,
719 (1969).
75. J. Pensky and H. G. Schwick, Science, 163, 698 (1969).
76. S. Ruddy, L. G. Hunsicker, and K. F. Austen, J.
Immunol., 108, 657 (1972).
77. J. Pensky, C. Hinz, E. Todd, R. Wedgewood, J. Boyer,
and I. Lepow, J. Immunol., 100, 142 (1968).
78. T. Boenisch and C. A. Alper, Biochem. Biophys. Acta,
221, 529 (1970).
79. R. A. Nelson, Jr. and C. Biro, Immunology, 14, 527
(1968).
80. H. S. Shin, H. Gewurz, and R. Snyderman, Proc. Soc.
Exptl. Biol. Med., 131, 203 (1969).
81. H. J. Müller-Eberhard and K. E. Fjellström, J. Immunol.,
107, 1666 (1971).
82. H. Randall, S. Talbot, H. Neu, and A. Osler, Immunology, 4, 388 (1961).

83. S. Boyden, J. Exptl. Med., 115, 453 (1962).
84. R. Snyderman, H. S. Shin, J. K. Phillips, H. Gewurz, and S. E. Mergenhagen, J. Immunol., 103, 413 (1969).
85. R. A. Clark and H. R. Kimball, J. Clin. Invest., 50, 2645 (1971).
86. R. Snyderman, H. S. Shin, and M. H. Hausman, Proc. Soc. Exptl. Biol. Med., 138, 387 (1971).
87. R. Snyderman, J. K. Phillips, and S. E. Mergenhagen, J. Exptl. Med., 134, 1114 (1971).
88. P. A. Ward, Am. J. Pathol., 54, 121 (1969).
89. A. B. Kay, Clin. Exptl. Immunol., 7, 723 (1970).
90. R. Snyderman, L. C. Altman, M. S. Hausman, and S. E. Mergenhagen, J. Immunol., 108, 857 (1972).
91. H. U. Keller and E. Sorkin, Intern. Arch. Allergy Appl. Immunol., 31, 505 (1967).
92. P. A. Ward, C. D. Offen, and J. R. Montgomery, Federation Proc., 30, 1721 (1971).
93. R. B. Johnston, Jr., M. R. Klemperer, C. A. Alper, and F. S. Rosen, J. Exptl. Med., 129, 1275 (1969).
94. H. S. Shin, M. R. Smith, and W. B. Wood, Jr., J. Exptl. Med., 130, 1229 (1969).
95. J. H. Humphrey, R. R. Dourmashkin, and S. N. Payne, in 5th Intern. Symp. Immunopathol. (P. A. Miescher and P. Grabar, eds.), Schwabe, Basel, 1968, p. 209.
96. D. H. Kay (ed.), Techniques for Electron Microscopy, Blackwell, Oxford, England, 1965.
97. J. E. Mellma, E. F. J. Van Bruggan, and M. Gruber, J. Mol. Biol., 31, 75 (1968).
98. R. C. Valentine, B. M. Shapiro, and E. R. Stadtman, Biochemistry, 7, 2143 (1968).
99. R. R. Dourmashkin, G. Virella, and R. M. E. Parkhouse, J. Mol. Biol., 56, 207 (1971).
100. E. Shelton, K. Yonemasu, and R. M. Stroud, Proc. Natl. Acad. Sci., U.S., 69, 65 (1972).
101. J. H. Humphrey and R. R. Dourmashkin, Adv. Immunol., 11, 75 (1969).
102. M. M. Frank, R. R. Dourmashkin, and J. H. Humphrey, J. Immunol., 104, 1502 (1970).
103. F. A. Medhurst, A. A. Glynn, and R. R. Dourmashkin, Immunology, 20, 441 (1971).
104. A. Klug, Phil. Trans. Roy. Soc. London, B261, 173 (1971).
105. S. E. Svehag and B. Bloth, Acta Pathol. Microbiol. Scand. (B), 78, 260 (1970).
106. M. J. Polley, in Progress in Immunology (B. Amos, ed.), Academic Press, New York, 1971, p. 597.

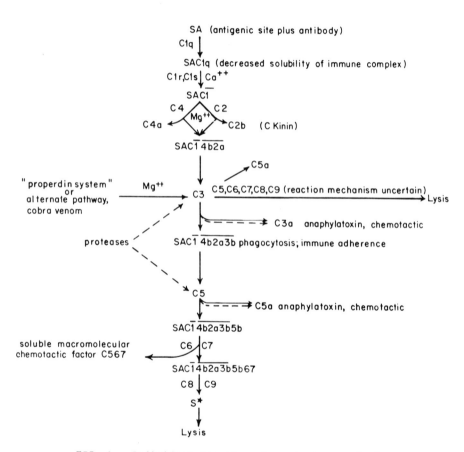

FIG. 1.   Individual reaction steps in immune lysis.

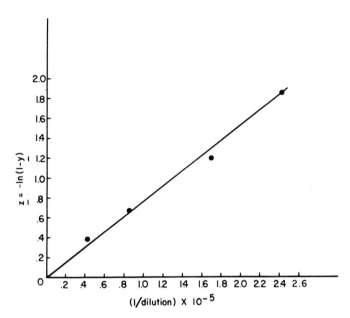

FIG. 2. Represented are the dose-response data from an
actual Cl titration. The z value is the $-\ln(1 - y)$, where y is
the percentage of cells hemolyzed. The ordinate is the decimal
equivalent of the dilution fraction (e.g., $1/100{,}000 = 1 \times 10^{-5}$).
The dilution that gives a z value of 1.00 in this example is
$1.35 \times 10^{-5} = 1/74{,}000$, and $74{,}000 \times 1.54 \times 10^8/ml = 1.1 \times 10^{13}$
effective Cl molecules/ml.

Chapter 8

ISOLATION OF MACROPHAGE POPULATIONS

Joseph C. Pisano

Department of Physiology
Tulane University School of Medicine
New Orleans, Louisiana

I.   HEPATIC MACROPHAGES 213

II. MACROPHAGES OF SPLEEN 215

III. PULMONARY AND PERITONEAL MACROPHAGES 217

IV. LYMPHOCYTES 217

     REFERENCES 218

I. HEPATIC MACROPHAGES[1]

Isolation of hepatic macrophages was first described nearly

40 years ago [3]. This early method capitalized upon the great

---

[1]Functions and characteristics of macrophages, as well as an
extensive bibliography, can be found in monographs by Pearsall
and Weiser [1] and by Cohn [2].

propensity of macrophages to engulf particulate material.
Powdered iron was injected into an animal; then, after an adequate period of time during which phagocytosis could occur,
livers were removed and a dispersed suspension of liver cells
was prepared. The suspension of cells was placed in a magnetic
field and the iron-containing macrophages were collected by
decanting cells without iron. This method yielded a reasonably
pure population of macrophages; however, the physiological con-
sequences of engulfed iron particles has never been adequately
assessed. Our own experiences with this procedure have always
yielded cells that manifest much blebbing of the membrane and
only slight phagocytic activity.

More recently, enzymatic methods for the isolation of
hepatic macrophages have been described [4,5]. These "updated"
procedures call for the digestion of liver tissue with enzyme
solution. A relatively pure population of hepatic macrophages
is obtained, with a higher degree of phagocytic activity than
is the case in cells obtained by the use of powdered iron [3].
However, the effects of the enzymes used (collagenase, hyaluro-
nidase, trypsin, and DNase) on the integrity of the isolated
cells have not been described. In addition, these procedures
require 4-8 hr for complete isolation of the macrophages. Cells
isolated from enzymatic digests of tissue by a prolonged procedure cannot be expected to remain in an unaltered physiological state.

A simple and rapid procedure for the isolation of hepatic macrophages has been described [6]. Livers can be perfused with a dilute solution of EDTA and macrophages are flushed out with the perfusate. Unfortunately, these macrophages lose the ability to phagocytize and cannot, therefore, be considered functionally normal. Thus, no adequate means exist for the isolation of a relatively homogeneous, physiologically normal population of hepatic macrophages.

II. MACROPHAGES OF SPLEEN

About the best available technique that now exists allows splenic macrophages to adsorb onto either glass or plastic surfaces [7]. Theoretically, splenic macrophages are the only cells within the spleen that will adhere to glass surfaces and, thus, simple decanting and washing removes from the surface all but the splenic macrophage population. Our experience with this technique has led to some serious questions about the purity of the cellular species isolated. If colloidal carbon particles are added to the adherent cell population, they do indeed exhibit active phagocytosis, suggesting that they are a population of macrophages. However, if colloidal carbon particles are added to the nonadherent cell population about 20% of the cells in this fraction will also engulf the carbon.

Whether these 20% of nonadherent cells represent macrophages
that have not adhered to the glass surface or whether they represent a nonadherent population of macrophages has not been elu-
cidated. Furthermore, if the adherent cells are isolated from
immune animals, they produce antibody. According to the description of the glass-adherent technique by Mosier [7], the adherent
cells should not possess this capacity to produce antibody;
indeed, the consensus of opinion is that macrophages do not
produce antibody. Thus, again we are faced with a problem; are
the adherent cells, which produce antibody, a population of
splenic lymphocytes that are capable of adhering to glass, or
are they macrophages that can produce antibody?

An additional difficulty with the method of glass adherence
for the isolation of splenic macrophages is the discrepancy
between the proportion of cells that adheres to the glass and
estimates of the number of macrophages that is present within
the total spleen-cell population. The highest estimate that we
have seen of splenic macrophages is between 25% and 33% of the
total nucleated cell population of the spleen. By the technique
of glass adherence, we have obtained as much as 50% of the total
spleen-cell population adhering to both glass and plastic surfaces. Other investigators have reported similar findings.
These difficulties suggest that the homogeneity of splenic
macrophages isolated by this procedure must be suspect.

III. PULMONARY AND PERITONEAL MACROPHAGES

Pulmonary and peritoneal macrophages are, by far, the simplest to obtain. Pulmonary macrophages can be isolated by simple pulmonary lavage [9]. However, this technique is not readily adaptable to small animals, such as rats and mice, because of the low yield of cells thus obtained. As with the pulmonary macrophage, peritoneal macrophages are obtained by lavage techniques. Again, a low yield of cells is obtained, but the yield can be improved (to as much as 100-fold) by stimulation of the peritoneal cavity 3-4 days before harvesting of macrophages, with agents such as sodium caseinate, glycogen, or starch. Unfortunately, cells from stimulated animals behave differently than those from normal, unstimulated animals in terms of enzyme activity [10], homing response [11], and immuno- logical activity [12]. In addition, the population of cells obtained from a peritoneal cavity is generally contaminated with significant numbers of lymphocytes and polymorphonuclear leucocytes.

IV. LYMPHOCYTES

Lymphocytes, which are a frequent contaminant of isolated macrophage preparations, have been described as possessing phagocytic capacity [13,14]. In addition, there are reports of transformation of lymphocytes into macrophages [15,16]; if this is the case, then the complete separation of these two cell types may not be possible.

REFERENCES

1. N. N. Pearsall and R. S. Weiser, The Macrophage, Lea and Febiger, Philadelphia, 1970.

2. Z. A. Cohn, Advan. Immunol., 9, 163 (1968).

3. P. Rous and J. W. Beard, J. Exptl. Med., 59, 577 (1934).

4. J. C. Pisano, J. P. Filkins, and N. R. Di Luzio. Proc. Soc. Exptl. Biol. Med., 128, 917 (1968).

5. W. G. Howard, G. H. Christie, J. L. Boak, and E. Evans-Anfolm, in La Greffe des Cellules Hematopoietiques Allogeniques, C.N.R.S., Paris, 1965, p. 95.

6. R. K. Fred, personal communication.

7. D. E. Mosier, Science, 158, 1573 (1967).

8. N. K. Jerne and A. A. Nordin, Science, 140, 405 (1963).

9. Q. M. Myrvik, E. S. Leake, and B. Fariss, J. Immunol., 86, 133 (1961).

10. R. Rossi and M. Zatti, J. Exptl. Med., 45, 548 (1964).

11. R. W. Gillette and E. M. Lance, J. Reticuloendothel. Soc., 10, 223 (1971).

12. R. J. North and G. B. Mackaness, Brit. J. Exptl. Pathol., 44, 608 (1963).

13. J. W. Rebuck, A. J. Petz, J. M. Riddle, R. H. Priest, and J. A. Logrippo, in Biological Activity of the Leucocyte, Ciba Foundation Study, Group No. 10, London, 1961.

14. M. W. Elves, in The Lymphocytes, Lloyd-Luke Ltd., London, 1966, p. 98.

15. J. W. Rebuck and J. H. Crowley, Ann. N. Y. Acad. Sci., 59, 757 (1955).

16. H. Braunsteiner, J. Partan, and N. Thumb, Blood, 13, 417 (1968).

AUTHOR INDEX

Numbers in brackets are reference numbers and indicate that an author's work is referred to although his name is not cited in the text. Underlined numbers give the page on which the complete reference is listed.

A

Agnello, V., 173[47], 207
Ahmed, M., 108[1], 134
Albertson, P., 24[18], 32
Allan, R., 171[41], 207
Allen, D.W., 119[2,3,68], 134, 137
Allen, J.M.V., 145[10], 160
Allison, A.C., 108[105], 116 [105], 117[105], 139
Alper, C.A., 183[78], 191 [93], 208, 209
Altman, L.C., 190[90], 209
Andres, G.A., 60[2], 81[14], 99, 100
Aoki, T., 119[32], 120[32], 135
Armstrong, W.D., 143[7], 159
Arroyave, C.M., 183[72], 208
Austen, K.F., 164[4], 166[20, 29], 167[30], 173[45], 183[76], 206, 207, 208
Avrameas, S., 80[6], 81[6], 95[6], 97[6], 98[6], 99[6,47], 100, 101
Axen, R., 29[22], 32

B

Bach, F., 129[4,5,6,46], 134, 136

Baker, P.J., 47[10], 48[10], 57
Banovitz, J., 171[40], 207
Barenboim, G.M., 80[2], 100
Bariety, J., 99[48], 101
Baril, E.F., 137
Barnett, E., 82[17], 88[17], 100
Barr, H.J., 80[4], 100
Bassin, R.H., 108[57], 136
Battips, D.M., 117[67], 137
Bauer, H., 108[7,8,12], 116 [7,8], 117[12], 135
Beard, J.W., 124[66], 213[3], 214[3], 137, 218
Becker, E.L., 166[24], 206
Bellon, B., 99[48], 101
Benjaminson, M.A., 81[16], 100
Benson, H.N., 43[5], 57
Beverley, P.C.L., 108[105], 116[105], 117[105], 139
Bienenstock, J., 91[28], 101
Billingham, R.E., 115[10], 129[9,10], 135
Biro, C.E., 183[69], 184[79], 208
Birry, A., 84[20], 94[39], 100, 101
Blackman, K., 108[13], 135
Bloom, B.B., 129[11], 135
Bloth, B., 204[105], 209
Boak, J.L., 214[5], 218
Bodmer, J., 80[5], 100
Bodmer, W., 80[5], 100
Boenisch, T., 183[78], 208
Bokisch, V.A., 164[9], 206

Bolognesi, D.P., 108[8,12],
    116[8], 117[12], 135
Bond, H.E., 177[58], 183[58],
    208
Boode, J.H., 92[32], 101
Boone, C.W., 108[13], 135
Borson, T., 129[19], 135
Borsos, T., 166[22,26,27], 167
    [31], 169[27], 171[38],
    173[52], 174[22,27,54,56],
    176[31], 177[58,59], 181
    [27,63], 183[58], 206,
    207, 208
Bosmann, H.B., 118[14], 135
Bova, D., 108[77], 137
Boyden, S.V., 115[15], 116
    [15], 186[83], 190[83],
    135, 209
Boyer, J., 183[77], 208
Boyse, E.A., 120[73], 129[73],
    137
Brade, V., 183[68], 208
Brandzaeg, H., 92[31], 101
Braunsteiner, H., 218[16], 218
Breese, Jr., S., 72[4,5,6],81
    [15], 116[16], 76, 100, 135
Brumfield, H.P., 43[5], 57
Brent, L., 129[9,17], 135
Bretton, R., 99[47], 101
Brodey, R.S., 108[31], 119[31],
    135
Brown, J.B., 129[17], 135
Brown, R.G., 117[81], 129[81],
    138
Bruchhausen, D., 87[23], 100
Brunette, D.M., 24[17], 32
Brusman, H.P., 87[25], 101
Budd, G.C., 81[10], 100
Burgin-Wolff, A., 94[38], 101
Burnet, F.M., 142[1], 159
Busch, G.J., 99[46], 101
Bussard, A., 41[3], 57

Cantrell, J., 165[14], 206
Capps, W.I., 108[41,44], 115[41],
    117[41], 119[41], 122[41],
    123[41], 136
Caroline, W.L., 108[105], 116
    [105], 117[105], 139
Carswell, E.A., 120[73], 129[73],
    137
Chan, M.M.W., 94[36], 101
Chaperon, E.A., 144[8], 160
Chase, M.W., 129[11,18], 135
Chatman, D., 165[13], 206
Cherry, M., 108[61B], 115[61B],
    116[61B], 117[61B], 137
Chiller, J.M., 155[13], 157[15],
    160
Christie, G.H., 143[4], 158[17],
    214[5], 218
Churchill, W.H., 129[19], 135
Claman, H.N., 144[8], 160
Clarke, D.A., 120[73], 129[73],
    137
Clark, H.F., 86[22], 100
Clark, R.A., 190[85], 209
Coe, J.E., 108[21], 116[21], 135
Cohen, A., 149[12], 160
Cohen, S., 172[44], 207
Cohn, Z.A., 213[2], 218
Coligan, J., 6[9], 31
Colten, H.R., 166[22], 171[38],
    174[22], 177[58], 183[58],
    206, 207, 208
Cook, C.T., 177[61], 181[63],
    183[61,68], 208
Cooper, N.R., 163[2], 205
Cornville, R., 50[22], 57
Courtenay, B.M., 143[4], 159
Cowan, K.M., 71[3], 76
Croce, C.M., 108[20], 115[20],
    135
Crowley, J.H., 218[15], 218
Cunningham, A.J., 49[11,12], 57

                C   D

Calcott, M.A., 164[7], 206   Dalmasso, A.P., 164[7], 206
Caloenescu, M., 94[39], 101   Dandliker, W.B., 30[23], 32

Davey, F. R., 99[46], 101
David, J.R., 116[22], 117[22], 129[22,23], 135
Davis, B.J., 182[64], 208
Davis, N.C., 166[19], 206
Day, E.D., 117[59], 137
deBracco, M.M.E., 173[50], 183 [50], 207
Deckers, P.J., 115[24], 135
deHaven, E., 108[31], 119[31], 135
Del Villano, B.C., 10[9a], 31
DeSaussure, V.A., 30[23], 32
Diener, E., 143[7], 159
DiLuzio, N.R., 214[4], 218
DiStefano, H.S., 108[25], 135
Dixon, F.J., 3[5,6], 4[5,8], 13[11a], 14[5,6], 52[26], 93[34], 157[16], 31, 57, 101, 160
Doebbler, T.K., 117[81], 129 [81], 138
Dolezel, J., 91[28], 101
Domanskii, A.N., 80[2], 100
Donaldson, V.H., 164[6], 206
Dorling, J., 99[44], 101
Dougherty, R.M., 108[25,62], 116[62], 135, 137
Dourmashkin, R.R., 194[95], 196[99], 199[99,101,102], 203[102,103], 205[101], 209
Drake, W.P., 116[26], 135
Dreskin, R.B., 96[41], 101
Dresser, D.W., 46[8], 57
Druet, P., 99[48], 101
Dube, O.L., 129[65], 137
Duesberg, P.H., 108[27], 135
Dulbecco, R., 115[101], 138
Dutton, R.W., 48[14], 129[28], 143[3,6], 154[3], 57, 135, 139

E

Eagle, H., 108[20], 115[20], 135

Eddy, B.E., 108[97B], 138
Edelman, G.M., 116[85], 117[85], 138
Edgington, T.S., 93[34], 101
Egan, M., 6[9], 48[16], 31, 57
Eisen, H., 117[40], 136
Elkind, M.M., 115[29], 135
Ellison, J.R., 80[4], 100
Elson, J., 158[17], 160
Elves, M.W., 218[14], 218
Ernback, S., 29[22], 32
Eskenazy, M., 48[18], 57
Evans-Anfolm, E., 214[5], 218

F

Fariss, B., 217[9], 218
Farquhar, M.G., 81[8], 100
Farr, R.S., 3[4], 31
Fefer, A., 117[60], 129[60], 137
Feldman, J.D., 52[23], 57
Feldman, M., 115[103], 117[103], 138
Feltkamp, T.E.W., 92[32], 101
Fenner, T., 142[1], 159
Fernandes, M.V., 129[54], 136
Filkins, J.P., 214[4], 218
Fink, M.A., 108[105], 116[105], 117[105], 139
Firschein, I.L., 129[46], 136
Fischinger, P.J., 108[30,57], 117[30], 119[30], 135, 136
Fisher, C.L., 119[74], 137
Fitzgerald, P.H., 129[69], 137
Fjellstrom, K.E., 184[81], 208
Fleischman, J.B., 94[37], 101
Ford, R., 165[15], 206
Foreman, C., 108[77], 119[75], 137
Frank, M.M., 171[39], 199[102], 203[102], 207, 209
Fred, R.K., 215[6], 218
Friedman, E., 164[8], 206
Fudenberg, H.H., 48[9], 57

G

Gamble, C.N., 156[14], 160
Gandini-Attardi, D., 94[37],
    101
Geering, G., 108[31], 119[31,
    32], 120[32], 135
Geery, G., 108[45], 136
Gell, P.G.H., 2[1], 31
George, M., 116[33], 117[33],
    129[33], 130[33], 136
Gewurz, H., 184[80], 186[84],
    188[84], 208, 209
Giebler, P., 117[34], 124[34],
    136
Gigli, I., 166[28], 167[28],
    171[37], 177[28], 178
    [28], 181[28], 183[28],
    191[37], 192[37], 206,
    207
Gilden, R.V., 108[35,77], 116
    [35,36], 119[35,36,74,
    75,76], 136, 137
Gillette, R.W., 217[11], 218
Glassock, R.J., 93[34], 101
Glick, M.C., 23[16], 24[16],
    32
Glynn, A.A., 203[103], 209
Gold, E.R., 48[9], 57
Goldenberg, D.M., 117[37], 136
Goldman, M., 81[11], 83[11],
    85[11], 87[11], 88[11],
    89[11], 100
Goldstein, D.A., 14[13], 32
Golub, E.S., 48[14], 143[3],
    154[3], 57, 159
Götze, O., 172[42], 183[42],
    207
Govaerts, A., 129[38], 136
Graf, T., 108[7], 116[7], 135
Graham, Jr., R.C., 81[9], 100
Green, M., 118[39], 136
Gruber, M., 195[97], 209

H

Hadding, R., 183[74], 208

Hannestad, K., 117[40], 136
Hansen, H.J., 117[37], 136
Hanstein, W.G., 13[11], 28[11],
    31
Hardy, W., 108[31,45], 119[31],
    135, 136
Harrell, B.E., 48[20], 57
Hartley, J.W., 108[41,44], 115
    [41], 117[41], 119[41],
    122[41], 123[41], 136
Hasham, N., 129[46], 136
Hatefi, Y., 13[11], 28[11], 31
Hausman, M.S., 190[86,90], 209
Heimer, G.V., 83[18], 100
Hellstrom, K.E., 106[42], 116
    [42], 129[42,52], 136
Hellstrom, I., 106[42], 116[42],
    129[42], 136
Henry, C., 117[50], 136
Herberman, R.B., 108[99,100A],
    116[99,100A], 117[99], 138
Hernandez, R., 94[38], 101
Herrick, C.A., 44[6], 57
Herzenberg, L.A., 116[47], 136
Hijmans, W., 92[30], 101
Hilgers, J., 108[45], 136
Hinz, C.F., 175[57], 183[77],
    208
Hirschhorn, K., 129[4,5,46],
    134, 136
Hodge, L.D., 14[12], 31
Hofermann, R., 87[23], 100
Hoffman, L.G., 169[33], 174[53],
    177[33], 207
Howard, J.G., 143[4], 158[17],
    159, 160
Howard, W.G., 214[5], 218
Howe, C., 60[1], 76
Hraba, T., 48[19], 57
Hsu, K.C., 60[1,2], 81[14], 99,
    100
Huebner, R.J., 108[27,35,41,43,
    44,61B], 115[41,43,61B,79],
    116[35,61B,80], 117[41,61B,
    80], 119[35,41,43,76,86],
    122[41], 123[41,43,79],
    135, 136, 137, 138
Humphrey, J.H., 194[95], 199[101,
    102], 203[102], 205[101],
    209

Hungerford, D.A., 117[67], 137
Hunsicker, L.G., 183[76], 208

I

Ingraham, J.S., 41[3], 57
Irvine, W.J., 94[36], 101
Ishizaka, K., 165[11], 171[11, 40], 206, 207
Ishizaka, T., 171[40], 207
Ishizaki, R., 108[100B], 123 [100B], 138
Isliker, H., 117[41], 207

J

Jackson, A.L., 48[17], 57
Jamieson, J.D., 116[49], 136
Jansen, I., 3[6], 14[6], 31
Jensen, J., 166[28], 167[28], 177[28], 178[28], 181 [28], 183[28], 206
Jerne, N.K., 41[2], 117[50], 143[2], 56, 136, 218
Johnson, G.D., 93[35], 99 [44], 101
Johnston, Jr., R.B., 191[93], 209
Jongsma, A.P.M., 92[30], 101
Julius, M.H., 116[47], 136
Just, M., 94[38], 101

K

Kahn, B.D., 28[20], 32
Kamen, M.D., 28[19], 32
Kao, M.S., 117[40], 136
Karnovsky, M.J., 81[9], 100
Kasatiya, S.S., 84[20], 94 [39], 100, 101
Kaufman, G.I., 95[40], 101
Kawamura, A., 83[13], 85[13], 87[13], 88[13], 89[13], 100

Kawarai, Y., 98[42], 101
Kay, A.B., 190[89], 209
Keller, H.U., 190[91], 209
Kelloff, G., 119[75], 137
Kennel, S.J., 14[13a], 21[13a], 28[19], 32
Kieler, J., 116[48], 117[48], 136
Kikuchi, K., 116[51], 136
Kikuchi, Y., 116[51], 136
Kimball, H.R., 190[85], 209
Kinsky, R.G., 158[17], 160
Kinsky, S.C., 166[25], 206
Kirschstein, R.L., 116[70], 137
Kirsten, W.H., 117[60], 129[60], 137
Kite, J.H., 117[81], 129[81], 138
Klassen, J., 91[29], 101
Klein, E., 129[52], 136
Klein, G., 129[52], 136
Klemperer, M.R., 191[93], 209
Klinman, N.R., 52[24], 57
Klug, A., 204[104], 209
Koffler, D., 93[33], 101
Koldovsky, P., 117[53], 129[53], 136
Kolodny, R.L., 129[46], 136
Koprowski, H., 108[20], 115[20], 129[54], 135, 136
Kraehenbuhl, J.B., 116[49], 136
Kretschmer, K.W., 164[6], 206
Kronman, B.S., 129[19], 135
Kunkel, H.G., 93[33], 173[46,47], 101, 207

L

Laliberte, F., 99[48], 101
Lance, E.M., 217[11], 218
Landy, M., 48[17], 57
Lane, W.T., 108[43,44], 115[43], 119[43], 123[43], 136
Lange, J., 108[30], 117[30], 119[30], 135
Last, J.A., 106[55], 108[55], 136

Laudenslayer, K., 177[61],
    183[61], 208
Lautenschleger, J., 6[9], 31
Lavrin, D.H., 108[99,100A],
    116[99,100A], 117[99],
    138
Lawrence, H.S., 129[56], 136
Leake, E.S., 217[9], 218
Lee, K.M., 108[57], 136
Lepow, I.H., 164[5,6], 165
    [17], 175[57], 183[77],
    206, 208
Lerner, R.A., 3[5,6,7], 4[5],
    14[5,6,12,13,13a], 21
    [13a], 31, 32
Levandoski, N., 30[23], 32
Levine, L., 115[58], 117[58],
    119[58], 137
Levy, N.L., 117[59], 137
Lewis, V.J., 119[78], 137
Linscott, W.D., 171[36], 190
    [36], 207
Livingston, D.M., 108[88],
    115[88], 116[88], 117[88],
    138
Logrippo, J.A., 218[13], 218
Lonai, P., 115[103], 117[103],
    138
Longmire, J., 3[4], 31
Loos, M., 174[54], 207
Louis, J., 157[15], 160
Lycette, R.R., 129[69], 137

M

Mackaness, G.B., 217[12], 218
Mage, R.G., 52[25], 57
Mahaley, M.J., 117[59], 137
Maizel, J.V., 17[14], 117[90],
    119[90], 32, 138
Mandel, L., 42[4], 57
Manni, J.A., 183[73], 208
Mardiney, M.R., 116[26], 135
Marquardt, H., 13[11a], 31
Martin, D.P., 116[70], 137
Martin-White, M.H., 108[77],
    137

Mason, T.E., 96[41], 101
Masuda, T., 116[47], 136
Maurer, P., 48[16], 57
Mayer, M.M., 163[1], 164[4,8],
    166[18,21,23,26], 168[32],
    169[32], 170[32], 171[32],
    172[32], 174[32], 177[60,
    61], 181[23], 183[23,60,
    61,68], 200[32], 205, 206,
    207, 208
Mayersbach, H.V., 87[23], 100
McCluskey, R.T., 91[29], 101
McCollum, W.H., 72[6], 76
McConahey, P.J., 3[5,6], 4[5,8],
    14[5,6], 52[26], 157[16],
    31, 57, 160
McCoy, J.L., 117[60], 129[60],
    137
McCoy, N.T., 117[60], 129[60],
    137
McKinney, R.M., 87[24], 101
McKenzie, A.T., 166[18], 206
Medawar, P.B., 129[9,17,61A],
    135, 137
Medhurst, F.A., 203[103], 209
Meier, H., 108[61B], 115[61B],
    116[61B], 117[61B], 119[98],
    137, 138
Meinke, W., 14[13], 32
Mellma, J.E., 195[97], 209
Mellman, W.J., 117[67], 137
Mellors, A., 164[8], 206
Merchant, B., 48[19,20], 57
Mergenhagen, S.E., 186[84], 188
    [84], 190[87,90], 209
Merler, E., 164[6], 206
Metzger, H., 2[3], 31
Meyers, P., 108[62], 116[62],
    137
Milgrom, F., 91[29], 101
Miller, J.A., 166[18,23], 181
    [23], 183[23], 206
Miller, J.F.A.P., 48[21], 148
    [11], 57, 160
Millette, C.F., 116[85], 117
    [85], 138
Mishell, R.I., 48[14], 143[3,6],
    154[3], 57, 159

Mitchison, N.A., 129[63,64,65], 137
Mommaerts, E.B., 124[66], 137
Montgomery, J.R., 190[92], 209
Moon, H.D., 116[104], 129[83], 138
Moore, J., 116[48], 117[48], 136
Moorhead, P.S., 117[67], 137
Morgan, C., 60[1], 76
Morrison, M., 18[15], 32
Morton, R.K., 13[10], 31
Mosier, D.E., 215[7], 216[7], 218
Müller-Eberhard, H.J., 163[2,3], 164[5,7,9], 165[16], 172 [42], 173[46], 178[62], 183[42,69,70,71,72,73, 74], 184[81], 205, 206 207, 208
Myers, D.D., 119[98], 138
Myrvik, Q.M., 217[9], 218

N

Naff, G.B., 175[57], 208
Nagaki, K., 174[55], 183[55], 207
Nakane, P.K., 98[42], 101
Nairn, R.C., 81[12], 83[12], 85[12], 87[12], 88[12], 89[12], 100
Natt, M.P., 44[6], 57
Nelson, R.A., 166[28], 167[28], 171[37], 177[28], 178[28], 181[28], 183[28,65,67], 184[79], 191[37], 192[37], 206, 207, 208
Nester, J.F., 95[40], 101
Neu, H., 184[82], 208
Niall, H.D., 119[3,68], 134, 137
Nilsson, U.R., 183[71], 208
Nishioka, K., 165[10], 171[36], 190[10,36], 205[10], 206
Nomura, S., 108[57], 136

Nordin, A.A., 41[2], 117[50], 143[2], 56, 136, 218
North, R.J., 217[12], 218
Nossal, G.J.V., 41[1], 56
Nowell, P.C., 117[67], 137
Nowinski, R.C., 108[45], 136

O

Oettgen, H.F., 120[73], 129[73], 137
Offen, C.D., 190[92], 209
Old, L.J., 108[31], 119[31,32], 120[32,73], 129[73], 135, 137
Oliveira, B., 172[43], 207
Oroszlan, S., 108[35,77], 116 [35,36], 119[35,36,74,75, 76], 136, 137
Osborne, M., 29[21], 117[102], 119[102], 32, 138
Osler, A., 172[43], 184[82], 207, 208

P

Painter, J.C., 124[66], 137
Palade, G.E., 81[8], 100
Park, W.P., 108[87,88], 115[88], 116[87,88], 117[87,88], 138
Parkhouse, R.M.E., 196[99], 199 [99], 209
Partan, J., 218[16], 218
Paterson, P.Y., 116[22], 117[22], 129[22], 135
Pauling, V.L., 166[19], 206
Payne, S.N., 194[95], 209
Pearmain, G., 129[69], 137
Pearsall, N.N., 213[1], 218
Pearse, A.G.E., 80[1], 100
Pearce, G.W., 87[24], 101
Pellegrino, M.A., 28[20], 32
Pensky, J., 175[57], 183[75,77], 208
Petricciani, J.C., 116[70], 137

Petrunov, B., 48[18], 57
Pettengill, O.S., 94[37], 101
Petts, V., 99[45], 101
Petz, A.J., 218[13], 218
Phifer, R.F., 96[41], 101
Phillips, D.R., 18[15], 32
Phillips, J.K., 186[84], 188
   [84], 190[87], 209
Phillips, M.E., 116[51], 136
Pick, E., 52[23], 57
Pilch, Y.H., 115[24], 135
Pisano, J.C., 214[4], 218
Plaut, A.G., 172[44], 207
Ploem, J.S., 83[19], 84[19],
   92[30], 101
Pollard, M., 108[71], 117[108],
   137, 139
Polley, M., 163[2], 170[35],
   183[70], 204[106], 208,
   209
Pomeroy, B.S., 43[5], 57
Porath, J., 29[22], 32
Post, R.S., 88[27], 101
Potter, V.R., 137
Prescott, B., 47[10], 48[10],
   57
Priest, R.H., 218[13], 218
Pruitt, K., 165[14], 206

               R

Radjikowski, C., 116[48], 117
   [48], 136
Raff, M.C., 144[9], 160
Randall, H., 184[82], 208
Rapp, H.J., 129[19], 166[22,26
   27], 167[31], 169[27],
   171[38], 173[52], 174[22,
   27,54,56], 176[31], 177
   [58,59], 181[27,63], 183
   [58], 135, 206, 207, 208
Rebuck, J.W., 218[13,15], 218
Reid, R.T., 3[4], 31
Reiff, A.E., 145[10], 160
Reisfeld, M.A., 28[20], 32
Reisfeld, R.A., 52[25], 119
   [78], 57, 137

Rejnek, J., 52[25], 57
Rhim, J.S., 115[79], 116[80],
   117[80], 123[79], 137, 138
Riddle, J.M., 218[13], 218
Říha, I., 46[7], 57
Ripps, C., 129[5], 134
Robinson, H.L., 108[27], 135
Robinson, W.S., 108[27], 135
Roblin, R., 115[101], 138
Roitt, I.M., 99[45], 101
Rommel, F.A., 170[34], 207
Rose, F.C., 108[82], 138
Rose, N.R., 117[81], 129[81],
   138
Rose, S.M., 108[82], 138
Rosen, F.S., 164[6], 206
Rosenau, W., 129[83], 138
Rosenblum, D., 165[13], 206
Rossi, R., 217[10], 218
Rous, P., 213[3], 214[3], 218
Rowe, D.S., 50[22], 57
Rowe, W.P., 108[41,43,44], 115
   [41,43], 117[41], 119[41,
   43], 122[41], 123[41,43],
   136
Rubin, H., 108[84], 116[94], 138
Ruddy, S., 166[20,29], 167[30],
   183[76], 206, 207, 208
Rutishauser, V., 116[85], 117
   [85], 138

               S

Sainte-Marie, G., 88[26], 101
Sandberg, A.L., 172[43], 207
Sander, G., 85[21], 100
Sanderson, R.P., 48[17], 57
Sarma, P.S., 119[3,86], 134, 138
Sauer, R., 119[3,68], 134, 137
Schafer, W., 108[30], 117[30],
   119[30], 135
Schidlovsky, G., 108[1], 134
Schlesinger, M., 149[12], 160
Schreiber, A.D., 171[39], 207
Schreibman, R.R., 129[5], 134
Schuit, H.R.E., 92[30], 101
Schur, P.H., 93[33], 173[45],
   101, 207

Schwick, H.G., 183[75], 208
Scolnick, E.M., 108[87,88],
    115[88], 116[87,88], 117
    [87,88], 138
Segal, B.C., 60[2], 81[14],
    76, 100
Segre, D., 48[13], 57
Segre, M., 48[13], 57
Seligman, A.M., 81[7], 100
Sell, S., 2[1], 31
Sever, J.T., 119[89], 138
Shapiro, A.I., 17[14], 32
Shapiro, A.L., 117[90], 119[90],
    138
Shapiro, B.M., 196[98], 209
Sharp, D.G., 124[66], 137
Shelton, E., 199[100], 204
    [100], 209
Shepard, C.C., 86[22], 100
Shimada, A., 183[66], 208
Shin, H.S., 164[8], 165[15,23],
    177[60,61], 181[23], 183
    [23,60,61,68], 184[80],
    186[84], 188[84], 190[86],
    192[94], 206, 208, 209
Shipley, W.V., 115[91], 138
Shiu, G., 108[100A], 116[100A],
    138
Shnitka, T.K., 81[7], 100
Shope, R.E., 108[92], 138
Sibel, L.R., 108[105], 116
    [105], 117[105], 139
Silver, H., 48[21], 57
Silvers, W.K., 115[10], 129
    [10], 135
Sjogren, H.O., 129[52], 136
Sjoquist, J., 165[12], 206
Smith, J.K., 166[24], 206
Smith, M.E., 99[44], 101
Smith, M.R., 192[94], 209
Smith, R., 3[4], 31
Snell, G.D., 129[93], 138
Snyderman, R., 164[8], 184[80],
    186[84], 188[84], 190[86,
    87,90], 206, 208, 209
Sorenson, G.D., 94[37], 101
Sorkin, E., 190[91], 209
Southam, C.M., 116[51], 136
Spicer, S.S., 96[41], 101

Spillane, J.T., 87[24], 101
Spitzer, R.E., 166[19], 206
Stadtman, E.R., 196[98], 209
Stalenhein, G., 165[12], 206
Stanley, T.B., 119[74], 137
Stashak, P.W., 47[10], 48[10],
    57
Steck, F.T., 116[94], 138
Steel, C.M., 108[95], 116[95],
    138
Stenback, W.A., 108[96,97A],
    120[96,97A], 138
Šterzl, J., 42[4], 46[7], 57
Steuland, N.P., 169[33], 177[33],
    207
Stewart, S.E., 108[97B], 138
Stitzel, A.E., 166[19], 206
Stolfi, R., 170[34], 207
Strauss, W., 98[43], 101
Stroud, R.M., 164[4], 165[14],
    166[18], 173[48,49,50,51],
    174[55], 183[48,50,55],
    199[100], 206, 207, 209
Sutton, H., 115[29], 135
Svehag, S.E., 204[105], 209
Swallow, R.A., 96[41], 101
Sweet, G.H., 48[15], 57
Szenberg, A., 49[12], 57

                    T

Tado, A., 119[32], 120[32], 135
Takemoto, K.K., 108[21,99], 116
    [21,99], 117[99], 135, 138
Talbot, S., 184[82], 208
Tamura, N., 166[28], 167[28],
    177[28], 178[28], 181[28],
    183[28,65,66,67], 206, 208
Taylor, B.A., 108[61B], 115[61B],
    116[61B], 117[61B], 119[98],
    137, 138
Taylor, C.E.D., 83[18], 100
Taylor, R.B., 52[24], 57
Terynyck, T., 99[47], 101
Thumb, N., 218[16], 218
Till, J.F., 24[17], 32
Ting, C.C., 108[99,100A], 116
    [99,100A], 117[99], 138

Ting, R.C., 108[99], 116[99],
    117[99], 138
Todd, C., 6[9], 31
Todd, E.W., 175[57], 183[77],
    208
Tomasi, Jr., T.B., 172[44], 207
Toni, R., 108[77], 137
Toy, S.T., 108[105], 116[105],
    117[105], 139
Trenton, J.J., 108[96,97A],
    120[96,97A], 138
Triplett, R.F., 144[8], 160
Tripp, M., 80[5], 100
Turner, H.D., 108[27,43], 115
    [43,79], 119[43,86], 123
    [43,79], 135, 136, 137,
    138
Turoverov, K.K., 80[2], 100

U

Ulrich, K., 116[48], 117[48],
    136
Ungaro, P.C., 116[26], 135

V

Valentine, R.C., 196[98], 209
Van Bruggan, E.F.J., 195[97],
    209
Van Hoosier, G.L., 108[96,97A],
    120[96,97A], 138
Vaughan, J.H., 116[33], 117[33],
    129[33], 130[33], 136
Vinuela, E., 117[90], 119[90],
    138
Vinuela, F., 17[14], 32
Virella, G., 196[99], 199[99],
    209
Vogt, P.K., 108[100B], 123
    [100B], 138
Volanakis, J.E., 173[49], 207

W

Wallace, R.E., 116[70], 137
Walsh, P., 48[16], 57
Walter, G., 115[101], 138
Walters, C.S., 2[2], 31
Ward, P.A., 190[88,92], 209
Warner, N.L., 48[21], 57
Warren, L., 23[16], 24[16], 32
Wasserman, D.E., 95[40], 101
Watanabe, M., 137
Webb, J.A., 99[44], 101
Weber, J., 29[21], 32
Weber, K., 117[102], 119[102],
    138
Wedgewood, R., 183[77], 208
Weigle, W.O., 48[14], 143[3],
    154[3], 155[13], 157[15],
    57, 159, 160
Weiser, R.S., 213[1], 218
Welborn, F.L., 48[15], 57
Welcerle, H., 115[103], 117[103],
    138
Werner, R., 116[104], 138
West, C.D., 166[19], 206
Wheelock, E.F., 108[105], 116
    [105], 117[105], 139
Whittle, E.D., 137
Wigzell, J., 2[2], 31
Williams, D.E., 119[78], 137
Williams, L.B., 115[79], 123[79],
    137
Williamson, W.G., 94[36], 101
Willoughby, W., 165[15], 206
Wilson, D.B., 115[10], 129[10],
    135
Winchester, R.J., 173[47], 207
Winn, H.J., 115[106], 117[107],
    129[106,107], 139
Wissig, S.L., 81[8], 100
Wood, Jr., W.B., 192[94], 209
Wortis, H.H., 46[8], 57

Y

Yachnin, S., 165[13], 206
Yamada, A., 143[5], 159
Yamada, H., 143[5], 159
Yonemasu, K., 173[48], 183[48],
    199[100], 204[100], 207,
    209

Z

Zacharia, T.P., 106[55], 108
    [55,109], 116[16], 117
    [108,109], 135, 136, 139
Zacharius, R.M., 117[110], 127
    [110], 139
Zatti, M., 217[10], 218
Zell, E., 117[110], 127[110],
    139

SUBJECT INDEX

A

ABC-inhibition technique
  sensitivity, 13
Adenovirus, 105
  type 12, 106, 133
  type 19, 133
Agar-gel diffusion, 71
Alternate pathway to C3
  cleavage, 172
Analytical acrylamide gel
  electrophoresis, 127
Anaphylatoxin activity, 184
Anti-Θ antiserum, 144
Anti-Θ serum, 149
Antibodies of the IgG class, 46
Antibody, 81
  hybrid, 97
Antibody immobilized on a
  solid support, 29
Antigen, 81
Antigen-binding capacity test, 3
Antigen-binding cells, 157
Antigens; see also Group-
  specific antigens
  envelope, 118
  humoral, 106
  serological, 106
  thymus-dependent, 144
  thymus-independent, 144
  transplantation, 106, 109,
  129
Antigens conjugated to the SRBC
  membranes, 47
Antiserum and complement lysis,
  129
  technique, 132
Antiserum to group-specific
  antigens, 128

B

B cells, 148
Balanced salt solution, 145
Bone marrow, 91, 147
Boyden chambers, modified, 187
Bridge method, 97

C

C nomenclature, 163
C1 assay, 174
C1 fixation and transfer, 173
C1q
  precipitation of aggregated
    immunoglobulins, 173
  purified, 201
  structure of, 194
C2, purification procedure of,
  181
C3, isolation of, 177
CH50 assay, 172
Carbon-arc lamp, 83
Carbon films, 196
Cell line 8866, 14
Cell line WIL$_2$, 11, 14
Cellular intermediate, 170
Chick-embryo fibroblasts, 124
Cluster technique, 50

A (right column top)

Artificial substrate, 92
Autofluorescence, 83
Automated indirect immunofluo-
  rescence tests, 94
Autoradiography, 81
Avian leukosis, 105

231

Color photomicrography, 94
Combined fluorescence, 93
Complement components, refer-
    ences for the purification
    of, 183
Complement fixation, 90, 128
Complement fixation tests, 109
Complement reactivity, one-
    hit theory of, 166
Congo red, 93
Conjugation, 67
    of serum proteins to
    erythrocytes, 154
Conjugation buffer, 155
Counterstaining, 93
Cross-reacting antibody, 192
Cytochrome c, 81
Cytophilic antibody, 129
    technique, 131
Cytotoxic test, 111
    in vitro, 113

D

D-9 lymphosarcoma, 120
Dark-field condenser, 84
    cardioid type, 84
    reflecting type, 84
Delayed hypersensitivity, 129
Density-gradient centrifuga-
    tion, 120
Dichroic mirrors, 84
Diploid human lymphocytes, 3
Direct immunologic precipita-
    tion, 17
Dissolution procedures, 27
Dyes, 80

E

EAC4 cell, 174
Electron microscope, 60
Electron microscopy, 25, 81
Electrophoresis, 125; see also
    Preparative gel electro-
    phoresis

Elution, 93
Envelope antigens, see Antigens
Enzymes, 81
Epi-illumination, 84
Equine arteritis virus, see
    Viruses
Evans blue, 93

F

Fc portion, 171
Feline leukemia virus, see
    Viruses
Feline leukosis, 105
Ferritin, 60, 61, 81
Fetal-calf-serum proteins, 22
Filters, 83
    interference, 83
Fixatives, 89
Fluorescein diacetate, 80
Fluorescein isothiocyanate, 81
Fluorescein-to-protein (F/P)
    ratio, 87
Forssman antibody and guinea
    pig C, 199

G

Gel diffusion, 128
Gel staining, 127
Globulin, 65
Group-specific antigen of
    Rauscher leukemia virus, 122
Group-specific antigens, 106,
    118, 120, 133
    gs-1, 120
    gs-3, 120

H

Hamster leukemia virus, see
    Viruses

Hemagglutination, 109
    indirect, 109
    inhibition, 109
Hepatic macrophages, 213
Herpesvirus, see Viruses
High-pressure mercury lamp, 83
Histochemical methods, 80
Horseradish peroxidase, 81
Human-milk virus, see Viruses
Human PMN, 188
Human wart virus, see Viruses
Humoral antigens, see Antigens
Hybrid antibody, see Antibody

I

Immune adherence, 190
Immunoabsorbents, 95
Immunoenzyme techniques, 81
Immunoferritin, 60
Immunofluorescence, 109
    indirect, 109
Immunofluorescence adsorption
    test, 111
Immunogen, 156
Immunoperoxidase technique, 99
Incident light, 84
Interference filters, 83;
    see also Filters
Isolation and chemical
    characterization, 13
Isotopic antiglobulin, 129
    technique, 130

L

Lasers, 94
Lactoperoxidase system of
    cell surface iodination,
    15
Lesions produced by interac-
    tion of antibody and C,
    205
Leukemia-producing units, 113
Leukocyte chemotaxis assay,
    186

Light microscopy, 81
Liver (acetone) powder, 87
Localized immune lysis, 41

M

Mackaness-type chambers, 130
Macrophages of spleen, 215;
    see also Spleen
Membrane-associated immuno-
    globulins, 2
Membrane isolation, 15
Microcomplement fixation test,
    123
Microcytotoxicity test, 112
Moloney sarcoma virus, see
    Viruses
Monolayer method, 49
Mouse leukemia virus, see
    Viruses
Mouse sarcoma virus, Kirsten
    strain, see Viruses
Murine leukemia, 105
Murine leukemia virus, see
    Viruses

N

Negative staining, 194
Negative-staining membranes and
    molecules, 202
Negative stains, 197
Nonspecific immunoglobulins, 52
Novikoff ascites hepatoma, 120
Novikoff rat virus, see Viruses

O

Ouchterlony immunodiffusion
    test, 120

P

Papillomavirus, see Viruses
Peritoneal macrophages, 217
Phagocytic activity, 191
Phase contrast, 85
Phase-contrast microscopy, 131
Phosphatases, 81
Physical association of immuno-
  globulin with membrane
  fragments, 15
Plaque formation, 41
Plaque-forming cell, 41
Plaque-forming-cell (PFC)
  assay, 152
Polyomavirus, see Viruses
Preparative gel electrophore-
  sis, 126
Protein separation, 16
Pulmonary macrophages, 217

Q

Quantitation of membrane-
  associated immunoglobulin,
  3
Quinacrine hydrochloride, 80

R

Rabbit neutrophils, 187
Rat leukemia virus, see Viruses
Rauscher virus antigens, 123;
  see also Group-specific
  antigen of Rauscher leukemia virus
Rosette formation, 132
Rosette-forming cells (RFC),
  158

S

Sarcoma viruses, see Viruses

Serological antigens, see
  Antigens
Simian virus (SV40), see Viruses
Specificity, 91
Spleen, 146
  macrophages of, 215

T

Θ, 144
$(TCID)_{50}$, 133
Tetramethyl rhodamine isothio-
  cyanate, 81
Thymectomy, 148
Thymus, 145
Thymus-dependent antigens, see
  Antigens
Thymus-independent antigens,
  see Antigens
Tolerogen, 156
Toric lens, 84
Tracers, in vivo, 99
Transplantation antigens, see
  Antigens
Trapping of nonspecific material,
  27
Tungsten-halogen lamp, 83
Two-phase aqueous polymer
  partition, 24

U

Ultrafiltration, 124
Ultrastructure of small mole-
  cules, interpretation of,
  203
Ultraviolet fluorescence, 80
Uranium, 81

V

Vertical illumination, 84
Virus, 72

Virus neutralization, 114
  in vitro, 114
  in vivo, 114
  leukemia-virus, 114
Viruses, 81
  equine arteritis, 72
  feline leukemia, 120, 122
  hamster leukemia, 120
  herpes, 105, 106
  human-milk, 106
  human wart, 106
  Moloney sarcoma, 128
  mouse leukemia, 121
  mouse sarcoma, Kirsten
    strain, 122
  murine leukemia, 123
  Novikoff rat, 120

  papilloma, 105, 106
  polyoma, 105, 106
  rat leukemia, 120
  sarcoma, 123
  Simian (SV40), 105, 106

                  W

WIL₂ cells, see Cell line WIL₂

                  X

Xenon-arc lamp, 83